D0805826

EAT
MORE
BETTER

How to Make Every Bite
More Delicious

DAN PASHMAN
Creator and host of *The Sporkful*

Illustrations by Alex Eben Meyer

SIMON & SCHUSTER

NEW YORK   LONDON   TORONTO   SYDNEY   NEW DELHI   BRUNCH

Simon & Schuster
1230 Avenue of the Americas
New York, NY 10020

First Simon & Schuster hardcover edition October 2014

SIMON & SCHUSTER and colophon are registered trademarks of Simon & Schuster, Inc.

For information about special discounts for bulk purchases, please contact Simon & Schuster
Special Sales at 1-866-506-1949 or business@simonandschuster.com.

The Simon & Schuster Speakers Bureau can bring authors to your live event. For more information
or to book an event, contact the Simon & Schuster Speakers Bureau at 1-866-248-3049 or visit
our website at www.simonspeakers.com.

Interior design by Timothy Shaner, NightandDayDesign.biz

Illustrations by Alex Eben Meyer

Jacket design by Marlyn Dantes

Jacket illustration by Alex Eben Meyer

Manufactured in the United States of America

2 4 6 8 10 9 7 5 3 1

Library of Congress Cataloging-in-Publication Data is available upon request.

ISBN 978-1-4516-8973-0
ISBN 978-1-4516-8975-4 (ebook)

To my parents for their love and support,
and for sharing with me the joys of deliciousness

EAT
MORE
BETTER

SYLLABUS

The Quest for Perfect Deliciousness
A Mission Statement

I f you're like most people, you enjoy putting food in your mouth, chewing it, and swallowing it. But you probably don't derive as much pleasure from that pursuit as you could. This book will change that.

If a life contains a finite number of meals, and a meal contains a finite number of bites, you can only take so many bites before you're full and/or dead. A bite is a precious resource. It pains me to think of all the thoughtless eating that takes place across the world each day. So many mouthfuls meld together into one big, blah bolus we'll never get back. But let us not grieve for the bites that could have been. Let us instead look ahead, to those that are yet to be.

Every time you take a bite, you make important choices. When dipping a triangular tortilla chip, do you hold it by one point and dip two points, or hold it by a straight edge and dip one point? Your choice determines your maximum dip load, the optimal angle from which to enter the dip, and perhaps most importantly, the probability of chip breakage.

Make good decisions, and you'll get more pleasure from each bite. Multiply that increased enjoyment over years and decades, and the entire nature of your existence will change.

These choices *matter*. That's why every meal is an opportunity to achieve greatness, no matter how modest your culinary station. With training and study, you can *choose* to eat better, or as I like to say, Eat More Better—every bite, every meal, every day, for the rest of your life.

Taking the first step on this path requires that you understand yourself, and your purpose.

If you've managed to stay alive for more than a few days, which I gather you have, then you are an eater. But to reach life's mouthwatering mountaintop, you must also be an Eater: a seeker of the Platonic ideal known as Perfect Deliciousness.

Perfect Deliciousness is a gustatory nirvana, a higher state, greater even than the sum of all the sensory pleasures derived from eternally consuming the ideal bite. It's the Eater's true north as we forge through the dense jungle of meal mediocrity, slashing at the overgrown lettuce leaves of convention with our steak knives.

Most Eaters never get there. But it is the quest that defines us, and the earthly deliciousness we experience along the way that sustains us.

Of course, deliciousness is about much more than taste. As the Latin maxim goes, "*De gustibus non est disputandum.*" (In matters of taste, there can be no disputes.) That's why Eaters don't legislate personal preference. We uncover truths that transcend any one palate or plate. We see questions where others never thought to look and find answers where others never dared to tread. For instance:

- When eating a dish with many components, should you seek bite consistency or bite variety?

- What are the ethics of cherry-picking your favorite ingredients from a snack mix?

- What role does surface-area-to-volume ratio play in selecting the right ice cube, enjoying fried food, and countless other eating scenarios?

- Is an open-faced sandwich actually a sandwich?

- Is a cheeseburger better when the cheese is on the bottom, closer to your tongue, to accentuate cheesy goodness?

- If the three criteria for judging all pasta shapes are forkability, sauceability, and toothsinkability, is ziti really the best shape for baked ziti?

- When does it make sense to pour the milk into a bowl before the cereal?

- Is sliced bread actually the worst thing since itself?

- What's the best way to get seconds at an all-you-can-eat buffet?

- What scented hand soap pairs best with ratatouille?

Surely you agree that these are among the most pressing questions of our time.

As we answer them, we Eaters share our knowledge with fellow travelers across the Eatscape, so that all our lives may be made more delicious. We do this to uphold the motto of our dear alma mater, Sporkful University: Masticate, Ruminate, Promulgate.

SU may not yet have dorms or lecture halls, but it's long had a home on that most egalitarian of campuses, the Internet—through my podcast and blog, *The Sporkful.* (Motto: "It's not for foodies, it's for Eaters.") This book is Sporkful University's first textbook. It includes not only answers to the above questions but also the distillation of my many years of research and scholarship, as well as field-tested methods for eating success.

Why should you listen to me? Well for one thing, as I mentioned, I have a podcast. And they don't give those to just anyone. But more importantly, I've done the research.

I've spent years in the trenches of wedding reception cocktail hours, refining strategies for hors d'oeuvres aggregation and buzz management. I've lost countless chip fragments in the deep end of the dip to discover that a scoop chip is better used inverted, as a dome, which makes it virtually unbreakable. And I've endured untold syrup-soaked pancakes to develop the Porklift, a bacon lattice structure that reduces soggage by elevating your pancakes and, in turn, your eating experience.

By reading this book and following the curriculum it sets forth, you are embarking on a quest. It offers rewards that will blow your tongue's mind, but it's also fraught with risk. Make no mistake—this journey will change you.

The cherished notions with which you were raised will be challenged and laid bare for their foolishness. The orthodoxies of your past will become the blasphemies of your present. Some of you will object. Others will go insane.

For those who survive, no meal will look or taste the same again. A cheeseburger will only look right-side up if it's upside down.

Now, if you haven't already cast this book down in horror, let me lay out some guidelines:

- Every one of the techniques and recipes here can be reproduced with standard tools and ingredients in the comfort of your home or local eatery. In other words, don't tell me you couldn't complete a lesson because your dog ate your *sous-vide* machine.

- Balance your course load however you like. If you want to go from the beginning of the book to the end, great. But if you jump around a bit, you won't suffer indigestion. I recommend you keep the book near your dining table for handy reference in the years to come.

- Every kitchen is just a lab in disguise. At all times you are encouraged to experiment with your own versions of these techniques and recipes, and adapt them to your liking. Indeed, the absence of precise measurements in most recipes is intended to encourage this sort of participatory learning. If you discover a major improvement in a concept, contact me immediately at dan@sporkful.com or 908-9-SPORK-9.

- Several dishes in this book include ingredients that are uncooked, undercooked, or prepared in a manner radically divergent from both the norm and most board of health regulations. Be advised that by reading the sentence you're reading right now, you agree to release and hold harmless me, Simon & Schuster, and the as-yet-nonexistent corporation known as Sporkful Omnimedia from any lawsuits or claims if something bad happens to you or anyone else while doing anything this book instructs you to do. (I told you there were risks.) Now that you've read that sentence, I cite the legal principle of "No Backsies" to point out that it is binding.

The door to a new world is open before you. The bites of your past cannot be rebitten, but the bites of your future have yet to be written. So read on, before one more square inch of your stomach is misallocated. Effective immediately, you are an Eater. With study and care, you can choose to make your world a more delicious place. You can share your newfound knowledge with others across the Eatscape. And together, we will all learn to EAT MORE BETTER!

DAN PASHMAN
BRUNCH

1

PHYSICAL SCIENCES
Eating on Earth

- Surface-area-to-volume ratio (SATVOR)

- Avoiding soggage

- The right ice cubes for you

- Arguments against sliced bread and grilled hot dogs, and in favor of coffee ice cubes

- The Temperature of Life

- The Window of Optimal Consumption

- Baked potatoes and the Splitter's Dilemma

- The Proximity Effect and the Cheeseburger with Cheese on the Bottom

- Eating while looking at the sky

As Carl Sagan said, "If you wish to make an apple pie from scratch, you must first invent the universe."

Viewed as a whole, our universe feels intimidating and unknowable. But when it's broken down into its component parts, patterns and principles emerge. As far away as the stars may seem, they share a system with our planet, which is connected to our atmosphere, which is connected to the air that surrounds us, which goes into our mouths.

Eating is itself a physical science. Like chemistry, physics, and astronomy, it involves interactions between objects and forces that range from microscopic to massive. Gain an understanding of physical science, and you'll see that many of the same forces that move heaven and Earth can move you closer to Perfect Deliciousness.

Surface-area-to-volume ratio affects whole planets in much the same way it affects fried chicken. Explosions in space make new stars like explosions in your kitchen make new popcorn. And just as condensation forms clouds that can rain on your parade, it forms tiny beads of water that can rain on your grilled cheese.

The properties of physical science dictate the properties of your food—crisp,* crunch, temperature, texture, and gooey goodness, to name a few. You can't change the fundamental forces of nature, but you can understand them and, more importantly, eat in accordance with them.

As you work toward that goal, know that nothing in the Eatscape should ever be assumed. Every meal is a universe created from scratch, another opportunity to experience a big bang in your mouth. To make new discoveries, we Eaters must look upon the world the same way great thinkers always have. No detail is too small to be considered, or reconsidered. But

---

* Crisp: A noun, like crispiness, only crispier

we must also find time every day to contemplate the big picture, to look to the heavens—or the pantry—and wonder.

This chapter will take its cue from both the spirit and the substance of Sagan's statement about apple pie. We'll begin by discussing principles of physics and chemistry that affect our whole universe, from planet to plate. Then we'll discuss the process of scientific discovery, as well as what and how to eat when you're in the midst of it.

SURFACE-AREA-TO-VOLUME RATIO

This may not seem like the sexiest place to start, but follow me.

Why are certain fried foods crispier? Why do some foods heat up or cool down faster than others? Why do some absorb the liquids or sauces in which they're placed more readily? Why do some breads go stale faster than others?

SATVOR unlocks these mysteries.

SATVOR EXPLAINED

Surface-area-to-volume ratio, or SATVOR, is essentially the ratio between an object's size and the amount of exterior it has exposed to its surroundings. A food with a high SATVOR has a lot of surface in relation to its volume, like a waffle fry, or a chicken nugget shaped like the state of Maine—with a coastline of crunchy crags and crevices (figure 1.1).

In the case of fried foods, more relative surface area means more crispy fried exterior. In general, it means the food will heat up, cool down, and absorb liquid faster, all because it's in greater contact with its surroundings, whether those surroundings are a hot saucepan, a cold freezer, or a bubbling fryolator.

Fried Chicken: When Higher SATVOR Is Better

These nuggets each have the same amount of chicken, but the one shaped like Maine (*right*) has more surface area, which means more exterior exposed to battering and deep-frying, which means more crispy fried goodness.

Fig. 1.1

A food with a low SATVOR has more volume and less surface area, like a meatball. It takes longer to heat but holds its temperature longer once heated. It takes longer to absorb surrounding liquids, which can be bad when you seek saturation but good when you want to ward off soggage.

Sometimes an Eater wants high SATVOR, other times low. It can be a matter of personal preference. For instance, when putting chili and cheese on fries, high-SATVOR fries (like waffle or shoestring) tend to start with more crisp but turn soggy faster. Low-SATVOR fries (like the steak variety) will be less crispy, but better able to maintain distinct potato identity in the face of an onslaught of liquids.

The choice is yours. The key is to understand how SATVOR affects your eating experience so you can make more delicious decisions.

Let's analyze more ways in which SATVOR affects the Eatscape, starting with the microscopic. Some of the greatest enemies of deliciousness are invisible to the naked eye, which makes them all the more insidious.

SLICED BREAD: THE WORST THING SINCE ITSELF

Every minute that fresh bread is exposed to air, it becomes more stale and less delicious. Bringing it home from the bakery is like bringing a gunshot victim into the ER. The clock is ticking, and death creeps ever closer.

From the second you return home you must act as decisively as a dashing TV doctor. Any bread that will not be consumed within hours should be placed on life support. Wrap it tightly in plastic wrap, store it in a resealable plastic bag, and put it immediately in the freezer.

Whether or not the bread is bound for this icy ICU, however, the best way to reduce air exposure is to avoid slicing it. Slicing bread increases its SATVOR exponentially, making it far more vulnerable to its surroundings and reducing its life span.

I'm not talking about typical, mass-produced supermarket loaves. Those are treated with enough space-age ad-ditives to achieve eternal life. I'm talking about fresh bread from the bakery, where they ask if you want your loaf machine-sliced upon purchase. Such bread should be acquired intact. This increases life span by reducing SATVOR and allows the Eater to alter each slice's thickness depending on the situation (figure 1.2, page 13). This second benefit explains why I prefer my method over freezing a sliced loaf, which does ward off staleness almost as well as freezing a whole loaf.

TIP SATVOR explains why a long, skinny bread goes stale much faster than a chubby, round one, even when both are unsliced. Boules > baguettes.

DEFROSTED BREAD

Remove loaf from freezer and unwrap it. Microwave it on regular setting for just long enough to defrost outermost portions of bread. Better that you have to saw through some frozen bread than that you defrost too aggressively. (Use a serrated bread knife to slice.)

Even though cutting through partially frozen bread can be hard, it also has its advantages. Because the knife necessarily moves slower, it's easier to ensure even slices at the desired thickness.

Rewrap unsliced bread and return to freezer. If slices are still cold or partially frozen, let them come to room temperature. If you just can't wait, toast them briefly on low or microwave them for just a few seconds. You now have the best thing since fresh bread.

TIP There are a few high-quality breads, such as dense, moist German ryes and naturally fermented rustic loaves, that actually improve by sitting on the counter for a few days. Ask your baker if you're unsure, but if you don't know that you're buying one of those breads, you probably aren't.

TIP My objection to slicing notwithstanding, I do recommend halving a large bread and freezing the two halves separately (or eating one half immediately). This way the second half isn't taken in and out of the freezer and microwaved each time you want a slice, which can degenerate the bread over time—though not as much as exposing it to air all day, of course.

A Strategy for Slice Variation

Freezing a round bread in two halves not only keeps half the bread in a deep freeze until you're ready for it, it also offers you more flexibility in the size of your slices.

If you freeze a round bread whole, your first slices can only be very small ones from the end.

When you freeze the bread in halves, you can take your slice from the middle or end, varying in size depending on your needs.

ICE CUBES: AS IMPORTANT AS THE DRINKS THEY CHILL

An ice cube's SATVOR determines how quickly it will melt, which makes it a crucial consideration in any beverage. High SATVOR means a lot of the ice surface is in contact with the beverage, which is warmer than the ice, so the ice melts quickly. Low SATVOR means the opposite.

Generally speaking you want low-SATVOR ice so it melts slowly, chilling your beverage

TIP Ice above the surface of your drink does not chill the liquid nearly as well as ice that melts directly into the drink, because ice above drink level is pulling heat from the air directly surrounding it. Keep ice below the surface.

just enough without watering it down too much. However, a lot of industrial ice machines at bars and restaurants put the priority on making ice fast, which requires high SATVOR. (Of course, ice that's made fast also melts fast.)

There are various ice trays available online that make cubes with a low SATVOR. Some more thoughtful bars use single large cubes, three inches across each face. A few opt for a large sphere, which technically has the lowest SATVOR of all. But it has practical disadvantages (see figure 1.3, "The Problems with the Ice Sphere").

The Problems with the Ice Sphere

Too little ice-to-booze contact.

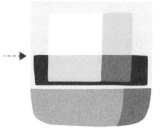

Fig. 1.3

The cube reaches the bottom of the glass—and the booze—more effectively.

Upon tippling, the sphere is more likely to touch the nose and face, an unpleasant sensation.

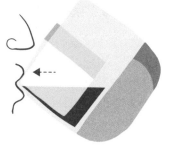

The cube rests comfortably against the upper lip.

THE ADVANTAGES OF MELTAGE AND DILUTION

Most cocktail recipes take ice meltage into account, making water a key component. When a bar uses low-SATVOR ice, the attendant dilution is part of the plan. On the other hand, when a bar uses high-SATVOR ice and the last few sips of the drink are mostly water, it usually means you got a bad drink. But there are times when more watering down is advantageous.

You may occasionally prefer a cold beverage that starts with a few strong hits of boozy punch, coffee zing, or lemonadey tart, then slowly mellows in strength and flavor. It's like rocketing into the sky, then parachuting gently back to Earth. Also, in especially hot weather, melted ice provides vital hydration.

THERMODYNAMICS & FOOD TEMPERATURE

The Zeroth Law of Thermodynamics makes it possible to measure temperature, but it's better known as the law that reminds us that "zeroth" is not a word. The other three laws of thermodynamics are even more useful, because the temperature at which a food is cooked and served has a huge impact on its taste, texture, mouthfeel, and more. That's why there are so many famous laws about it.

FIRST LAW: ADDING HEAT TO A SYSTEM
(or, Why Grilling Is Overrated)

The First Law of Thermodynamics states that if you add heat to a system, all the heat energy will either change the internal energy of the system, cause the system to do work, or both.

In other words, a grill generates heat in the form of fire. When people talk about grilling, that generates more heat, in the form of hot air. As more people grill and talk about how great grilling is, it makes grilling more popular (thus changing the system we call the Eatscape), and leads to more grilling (thus increasing work). This phenomenon is better known as Weber's Law.

Whole books have been written on grilling, and the very practice has become ensconced in a soot-black crust of romanticism. It conjures thoughts of summer, fun with friends and family, and a simpler time when food was cooked over an open flame and we didn't break bread with herbivores—we ate them.

Eaters are not swayed by such treacle.

To be clear, I'm an avid grillsman who enjoys scratching my Neanderthal itches as much as the next *Homo sapiens*, and I'm a strong supporter of experimentation and innovation. But *The Sporkful* is where sacred cows get grilled, and sometimes that means suggesting that sacred cows would be better prepared otherwise.

There's a notion in certain quarters that there's something inherently cool about grilling foods that are not traditionally grilled, as if grilling is a worthwhile end in itself. It is not.

In fact, our energy would be better spent identifying instances of grill overuse. We've been cooking over open flames for hundreds of thousands of years. It's far more likely that we continue to grill some foods out of mere habit than it is that grilling others never crossed our minds.

To wit . . .

DON'T GRILL HOT DOGS OR BUNS

Because hot dogs come precooked and contain enough fat to retain juice under the most adverse circumstances, and because most buns are awful to begin with, little thought is put into the preparation of these staples.

Grilling hot dogs is far more likely to dry them out, or at least leave them much less juicy than if you boil or steam them, which are better methods.

The only good argument in favor of the flame is that the process adds some crispy texture and char-grilled flavor. To that I would counter that hot dogs with a natural casing offer a unique textural sensation far preferable to that of a grill-induced crisp, and char flavor is better suited to foods with a lower SATVOR, where it's less apt to overpower.

The argument in favor of grilling buns is even less persuasive.

Most are so full of air that they lose their heat on the way from the grill to the mouth, and whatever crisp you attain comes with a dryness that Eaters cannot abide. (The top rack on most grills seems especially good at drying out bread without toasting it.)

If you want buns to be toasted, griddle them in a pan on the stove, with plenty of butter. This way you're adding moisture and flavor instead of sapping it, while still imparting texture and warmth. You can also steam buns by microwaving them in their plastic bag. Or make sun-steamed buns by leaving them in their bag in the sun. When moisture beads up on the inside of the bag, the buns are ready.

TIP Choose potato buns for hot dogs and burgers. They're more moist, flavorful, and sturdy than their standard-issue counterparts.

Have you noticed a pattern here? Neanderthals may have grilled a mean meat stick back in their day, but now we know that both hot dogs and buns are better when you cook them using a stove, indoors. That, dear pupils, is progress.

SECOND LAW: HEAT TRANSFER BETWEEN BODIES
(or, An Argument for Coffee Ice Cubes in Iced Coffee)

The Second Law of Thermodynamics states that it's impossible for a process to have as its sole result the transfer of heat from a cooler body to a hotter one without work also being done.

In other words, when ice is placed into coffee to make iced coffee, temperature transfer between the two bodies is not the sole result. The coffee is also watered down, unless you do work—by making your ice cubes out of coffee.

Iced coffee may be a wonderful, stiff elixir or a creamy sweet treat, depending on its preparation. However:

1. Lots of iced coffee is pretty bad, largely due to ice's meltability.

2. Bad iced coffee is a rip-off.

3. Iced coffee is often called "ice coffee," which is irksome.

All three of these problems can be solved by making iced coffee with coffee ice cubes. I'll address issue number one first.

In the early days of iced coffee, it was common to see purveyors simply pour hot coffee into a cup full of ice, causing all the ice to melt immediately and resulting in a beverage that was half coffee, half water. Horrific.

Nowadays more vendors recognize that coffee bound for ice must be chilled first, to minimize meltage. (Keeping it at room temperature is a poor alternative that remains disturbingly common.) The more thoughtful purveyors make a concentrated brew in anticipation of the watering down. Some have shifted to cold-brewed coffee, which has additional advantages. But these are convoluted solutions that ignore a simple one: Just make ice cubes out of coffee!

ICED COFFEE WITH COFFEE ICE CUBES

Make coffee and let come to room temperature. Put some in fridge and some in ice trays in freezer. When cubes are solid, combine with cold coffee from fridge. This way, the only thing that changes as the ice melts is the beverage's temperature, not its concentration.

This approach also alleviates concern number two, which is that a lot of iced coffee costs too much. It's one thing if a place is making coffee extra strong to counteract dilution, but if they're simply pouring their regular coffee over ice and charging a dollar more for it, it's a sham. If they used coffee ice cubes, the issue would be moot, because you'd actually be getting more coffee for the higher price, thanks to the larger cups used for iced coffee.

> TIP If the coffee part of the iced coffee you're buying is pre-chilled (as it should be) and made with regular ice, order it "without too much ice." The beverage will still stay cold for a while, and you'll get more coffee for your money.

As to concern number three, the other advantage of coffee ice cubes is that they offer a justification for the term "ice coffee." When you put regular ice into coffee, you are icing the coffee. As such, it is iced coffee. But when you freeze coffee, you change its state from liquid to solid, turning it into ice coffee. Thus, "ice coffee" is the appropriate term of art in the Eatscape for iced coffee with coffee ice cubes.

THIRD LAW: ENTROPY
(or, On the Temperature of Life)

Now that we've covered some vital interactions at the micro level of the glass and the plate, it's time to broaden our scope, by discussing the atmosphere that surrounds us.

"Room temperature" is a problematic term for Eaters, because it's imprecise. Depending on the HVAC system in question, it could mean two different temperatures in the same room. Furthermore, when you eat outdoors, it loses all meaning, since you're no longer in a room at all.

That's why we prefer to talk in terms of the Temperature of Life, which is the precise temperature of the air in the Eater's immediate vicinity at any moment.

The Third Law of Thermodynamics is "The entropy of a perfect crystal at absolute zero is exactly equal to zero." In other words, failure to eat in accordance with the Temperature of Life will result in entropy.

There are many advantages to serving and eating foods at life temperature. You don't have to worry about ensuring that different dishes are all at the right temperature at the right moment, which reduces kitchen stress. And when hosting, it allows you to prepare more food in advance and have more time with guests.

Many foods commonly served hot or cold are at least as good at life temperature. That being said, the goal is not to eat all foods at the Temperature of Life, but rather in accordance with it. At times, this means bucking conventional wisdom about the correct temperature at which foods should be served (see figure 1.4, "How the Quality of Certain Foods and Drinks Changes with Their Temperature").

How the Quality of Certain Foods and Drinks Changes with Their Temperature

This chart assumes a life temperature in the middle of the spectrum.

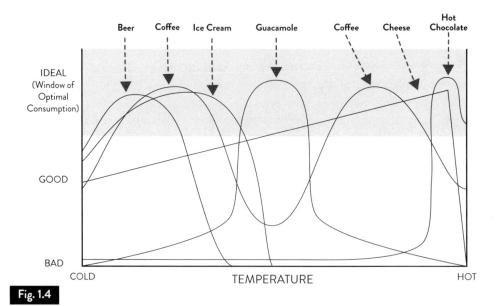

Fig. 1.4

COLD APPLE CRISP WITH WARM VANILLA ICE CREAM

The problem with the traditional way people serve the classic combo of hot pie and frozen ice cream is that it's an affront to the Temperature of Life. Fruit scalds your tongue and melts the ice cream so quickly that its Window of Optimal Consumption is barely one bite long. You do want ice cream to soften, because it has more flavor when it's not ice cold. You just don't want it to soften so fast. This recipe turns ice cream's Window of Optimal Consumption into a door.

Remove leftover apple crisp from fridge and ice cream from freezer. Scoop ice cream onto crisp and let sit until ice cream is about 25 percent melted.

ON THE PERILS OF CONDENSATION
(or, In Defense of Crisp)

While we're covering food temperature we must also cover condensation, a serious issue that comes into play when hot foods are transferred to cooler environs.

To crisp and crunch, condensation is Kryptonite. Any time a hot, crispy food is taken from where it was cooked and placed into an enclosed space, the steam it releases will condense, turn into water, and attack the food's toothsome exterior, bent on its soggification.

An enclosed space can be a container or bag, which is why, as discussed in the "Cultural Studies" chapter, French fries should never be taken to go or ordered for takeout, unless you plan to eat them on the way home. But other spaces are also problematic—interior regions of a pile of fried delicacies, or the area between a hot, crispy food and the plate on which it rests (see figure. 1.5, "Hot Grilled Cheese Under Attack from Condensation").

Hot Grilled Cheese Under Attack from Condensation

Fig. 1.5

Moisture eats away at crisp and turns bread soggy.

What to do? The easiest way to reduce condensation is to encourage air flow, so steam can escape without condensing. Use cookie racks to cool more than just cookies, and serve vulnerable sandwiches standing up (see figure. 1.6, "Vertical Sandwich Plating to Preserve Crisp").

Vertical Sandwich Plating to Preserve Crisp

Fig. 1.6

Because air can flow all around the sandwich, steam can escape from both sides and crispy grilled glory is preserved.

CONDENSATION AND BAKED POTATO TOPPING APPLICATION (THE SPLITTER'S DILEMMA)

The Splitter's Dilemma has long confounded physicists and Eaters alike. It goes like this:

When topping a baked potato, you want to maximize toppability, which means you want as much topping-to-potato contact as possible. If you only split the potato once, creating a V-shaped cavity, you'll end up with areas that are mostly topping and areas that are plain potato (figure 1.7). The alternative is to slice the potato lengthwise into wedges and essentially lay it out flat, to expose more potato interior to toppings (figure 1.8).

However, there is a competing concern. Wherever potato skin contacts plate, condensation will occur, and crisp will be lost. The more you split and pry open the potato to expose it to toppings, the more its skin necessarily touches the plate, and suffers as a result. (In other words, as potato SATVOR increases, both toppability and condensation increase as well.)

So what's the best approach?

It comes down to personal preference. Some Eaters (the "crispers") wouldn't dream of sacrificing crisp, and like having plain potato bites as palate cleansers. Others (the "toppers") want the focus on the interaction between potato and topping, and point out that real baked potato crisp is often short-lived anyway (though the skin's more subtle pop lives on).

Generally I'm a topper, but there's nothing wrong with being a crisper. The key is to pick a priority and seek to maximize it. Those who try to have it all are the real victims of the Splitter's Dilemma.

Traditional V-Cut Baked Potato for Increased Crisp

Large swaths of potato are far from the toppings, which is bad news for toppers. But less of the skin touches the plate, reducing condensation and preserving crisp—good news for crispers.

Bad for toppers.

Fig. 1.7

Good for crispers.

Wedge-Cut Baked Potato for Increased Toppability

There's much more potato-to-topping contact, which pleases toppers, but also much more skin-to-plate contact, which increases condensation and upsets crispers.

Good for toppers.

Fig. 1.8

Bad for crispers.

THE SCIENTIFIC METHOD: CURIOSITY, SERENDIPITY, DISCOVERY

One day, as the story goes, Isaac Newton was in the midst of an impromptu picnic when an apple fell on his head and inspired him to formulate his Law of Gravitation. Had he been eating indoors, we might still not know what keeps us stuck to this rock we call home. Instead, humanity experienced a great moment of discovery.

Pursuing Perfect Deliciousness means you're constantly curious, question everything, and hope that every once in a while, something delicious falls on your head. Of course, few things inspire curiosity more than looking at the sky and wondering, as Galileo and Copernicus did before us. And if you're like me, wondering makes you pretty hungry. (Perhaps being reminded of the gaps in our knowledge also reminds us of the emptiness in our stomachs.)

In this section I'll share some of my strategies for eating while stargazing, sky-gazing, or picnicking, so you can sustain yourself through your own process of scientific inquiry. In the end you'll see why Brillat-Savarin said, "The discovery of a new dish confers more happiness on humanity than the discovery of a new star."

THE PROXIMITY EFFECT AND PUTTING THE CHEESE ON THE BOTTOM

When you put any multifaceted food into your mouth, the components in closest proximity to your tongue are the ones you'll taste the most. This is the Proximity Effect, and it's a concept we'll return to throughout this book. Whether you're layering a food to be eaten by hand or preparing a taste on a fork, always consider how the bite will land on your tongue and which flavors you want to emphasize.

If you're picnicking or barbecuing, there's a good chance you're eating a sandwich or cheeseburger, and since we're only on chapter 1 here, there's a good chance it hasn't occurred to you to put the cheese on the bottom. If the sandwich or burger is already made, consider eating it upside down, to bring the cheese closer to your tongue and accentuate cheesy goodness. If not, here's a recipe:

CHEESEBURGER WITH CHEESE ON THE BOTTOM

Grill burgers and melt cheese on them on the grill as you normally would. Before burgers are done, lay out open buns nearby. As you remove burgers, place them on the top buns instead of the bottom ones. Put bottom buns on top and flip burgers over before serving. You just made cheeseburgers with cheese on the bottom. (Bonus: As burger juice flows downward, cheese seals it in and protects bottom bun from soggage.)

Many people say, "Why not just put the cheese on *both* sides? Then you'd have even *more* cheese!" True, but the burger is the star. You don't want to overshadow it, you just want to shine a little more of the spotlight on the cheese.

BETTER STARGAZING THROUGH BETTER S'MORES

Staring at the night sky always tastes better with s'mores. The problem is, s'mores have issues. Regulating the temperature of the marshmallow and chocolate can be difficult, but even worse, graham crackers are just a poor sandwich base. When you bite into one, the graham cracker usually splits and splinters, fillings are forced out the back, and the entire structure comes undone. (Issues of sandwichization are covered in far greater detail in chapter 3, "Engineering.")

I've done substantial research on s'mores, with the help of Nathaniel Goodyear, a president of the Sporkful Junior Eaters Society. (Every member is a president. After all, it's never too early to start accumulating extracurriculars for that college application.)

In order to improve s'mores, we must leave orthodoxy behind, in much the same way great astronomers have bucked convention to make major advances.

COPERNICAN HERESY S'MORES

Until now, Eaters assumed that graham crackers revolve around marshmallow and chocolate. But with Copernicus as my inspiration, I propose that in fact, the graham cracker is the center of the s'more universe. Not everyone will like this theory—until they taste it.

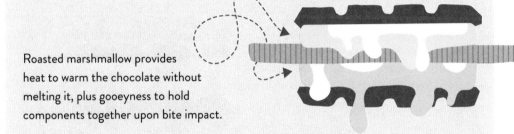

Roasted marshmallow provides heat to warm the chocolate without melting it, plus gooeyness to hold components together upon bite impact.

GALILEO'S GAMBIT S'MORES

This option replaces the graham cracker with an individual-size graham cracker pie crust, a move audacious enough to set off another Inquisition yet delicious enough to quell it. The pie crust's walls double as inclined planes, which make it easier to lift the food into your mouth. Put the chocolate on the pie crust and heat. Roast two marshmallows in the traditional way, then place them on top of the pie.

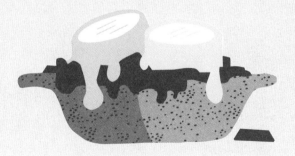

CHEWY CHIPS A-HAWKING S'MORES

When you bite into a typical s'more you get a messy explosion, with crumbs falling everywhere. But if you replace the graham crackers with soft, chewy cookies, the cookies will give to the bite and collapse into the black hole of your mouth, from which no food can escape.

Soft s'more exterior means teeth move effortlessly through the layers, so there's no pressure forcing fillings out the back.

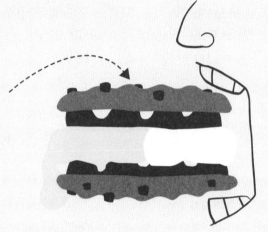

UNORTHODOX S'MORE INGREDIENTS

Whether you're sticking with a traditional s'more structure or trying one of the options in this chapter, you should experiment with unorthodox ingredients. Here are some options:

- peanut butter

- bacon

- thin-sliced apples

- dark chocolate instead of milk chocolate ("S'mores the Less Milky Way")

- chocolate-covered graham crackers—splintering is still a risk, but melting works better

POPCORN SUPERNOVA:
PROS AND CONS OF POPCORN SHAPES

In space, something is usually blowing up. But while these explosions are destructive, they also often create new stars. That's why one type of explosion is called a nova, Latin for "new."

Naturally, this leads us to a discussion of popcorn. Like stars, corn needs to explode to be reborn in the form of popcorn. And also like stars, popcorn has been around for a really long time. (Archaeologists found eighty-thousand-year-old fossilized popcorn beneath Mexico City.)

There's one more crucial similarity: Stars and popcorn kernels can take on many forms, but their shape is not random. With popcorn, it's largely determined by the type of kernel you begin with—mushroom kernels produce round balls while butterfly kernels produce pronged shapes.

I hasten to add, "mushroom" and "butterfly" are the proper terms used in traditional popcorn circles. But because we Eaters are more concerned with what to call them once they're popped (and thus edible), we use different terminology. A spherical popcorn piece produced by a mushroom kernel is called a Death Star. All the others, which tend to have smaller central orbs and various protrusions, are called Supernovae (singular: Supernova). Like this:

Death Star Kernel

Supernova Kernel

THE CRAB NEBULA: A NICE PLACE TO LOOK AT, BUT YOU WOULDN'T WANT TO EAT IT

A nebula in outer space may look stunning, but in reality, it's a cloud of gas and dust—a beautiful vision with little inside. It's no accident, then, that astronomers named the most famous nebula of all after the crab.

Whole blue crabs, well-known in the state of Maryland (among other regions), rank just above whole peanuts as the food that requires the most amount of work for the least amount of food. When the bill arrives, you realize you've just paid a lot of money to be left hungry, sore, and possibly bleeding, all so you could eat some crabs that are often seasoned so strongly you can't taste them anyway.

In what universe does it make sense to pay for such an experience? Lobster is labor-intensive and expensive, but at least it has toothsinkable pieces of meat to reward the Eater's efforts. It's one thing to cross the galaxy for a glass of water, it's another to do it for a sip.

ON PICNIC STYLE AND EFFECTIVE PICNICKING

Nighttime is not the only time to find inspiration in the sky. Picnicking is another great way to commune with nature and to eat in accordance with the Temperature of Life.

To picnic effectively, focus on what matters for picnic success: good weather, good friends, good food, and good drink. And because mobility is key, convenience is a major consideration. There are times when a more complicated setup requiring plates and utensils is appropriate (see "Extreme Picnicking," page 32). But those instances aside, ideal dishes will have the following attributes:

1. **EASY TO TRANSPORT FROM HOME TO PICNIC** Your setup is light and compact. All foods can be eaten from the same container or

wrapping in which they're transported. If you insist upon sipping your wine from something glass, just drink it straight from the bottle.

2. **EASY TO TRANSPORT FROM PICNIC TO MOUTH** All foods are either easily eaten by hand with little mess or already bite-sized. Dressings and sauces are minimal.

3. **DELICIOUS WHEN EATEN AT THE TEMPERATURE OF LIFE** You don't want to carry ice packs. You could bring a grill, but then it's not a picnic, it's a barbecue.

4. **PRODUCES MINIMAL MESS AND GARBAGE** Avoid plates and utensils by hosting a Toothpicnic (see below).

PICNICKING GLORY WITH THE TOOTHPICNIC

To execute a Toothpicnic, simply pre-cut everything from steaks to brownies into bite-sized morsels. Instead of sticking toothpicks into the foods before serving them, give one toothpick to each Eater to act as a primary utensil. This way each person can eat a whole meal with a single toothpick, reducing waste and eliminating the need for forks, knives, plates, and most napkins (because no food touches your hands). It's okay to provide more than one toothpick to Eaters concerned about inadvertently mixing flavors.

> **TIP** If you don't want to carry a wine opener, open the bottle before you leave and recork it. Or bring boxed wine—they say it's making a comeback, although I think it's been here for years.

> **TIP** You really don't want to bring ice packs. If you have kids, freeze juice boxes and use those to keep grown-up drinks cold. If you need more chilling power, put beer and wine on ice and use the ice to make cocktails. At least with ice you can dump it out almost anywhere, instead of having to carry it back home with you.

Pack foods in foil or reused takeout containers and transport everything in reused shopping bags, which double as garbage bags. If you execute your Toothpicnic well, you could even achieve that most elusive goal: the Empty-Handed Walkaway.

RANKING PICNIC FRUITS

How do these classic handheld fruits stack up for picnicking purposes?

1. **BANANA** The alpha fruit of the picnic, you can peel and eat it with virtually no mess. Bruising in transit is a concern, but if you start with a fairly firm one, you'll be fine.

2. **APPLE/PEAR (tie)** Apples are more durable and so better for transporting, but pears tend to be a little softer, and thus easier to bite into.

4. **ORANGE** If it's hard to peel you're in big trouble, and even if it's easy, you'll still end up sticky.

EXTREME PICNICKING

Any time you're transporting a meal to a remote location, you're raising the degree of difficulty. The natural reaction, then, is to keep things simple. But there's another way to look at it: The greater the challenge, the greater the glory to the Eater who conquers it.

So if you're going to bring a dish to a picnic that requires plates and utensils, bring many such dishes. Once you cross that line, you might as well take off running. Bring a nice blanket and a variety of chilled beverages, plus salt and pepper shakers. Serve a salad, transported with the dressing on the side so the greens don't sog.

In other words, you know the people in the catalog who have the wicker picnic basket with matching glassware and mobile wine cellar? Be those people. Make it an extreme picnic.

HOMEWORK

Understanding your physical world is crucial to eating more better, whether your primary concern is crisp, crunch, temperature, texture, gooey goodness, or any other property of your food. Once you've gained this understanding, the next step is to view each meal as a blank slate, another opportunity for an advance in the Eatscape.

Now here's your assignment . . .

Find a food not referenced in this chapter where SATVOR plays an important role in determining deliciousness. Alter its surface area and/or its volume to make its SATVOR more favorable. Then consider what other foods would benefit from the exact same treatment. You may be surprised when you realize how powerful a discovery you just made.

Submit your homework to me at dan@sporkful.com.

2

LANGUAGE ARTS
Better Communication, Better Consumption

- The strict-constructionist definition of a sandwich

- Why sparkling water is neither sparkling nor water

- A region's right to define a food it invents

- A screed against similes, metaphors, analogies, and oxymorons

- The difference between condiments, spreads, dips, and sauces

- Salad grammar, editing, legibility, and inversion

- Classic works of poetry and composition

Food is the universal language, like Esperanto, but useful. It is a form of self-expression and a means to communicate with others. To use it effectively you must understand its rules—and know when to break them.

At many points in this book, I encourage you to make eating choices based on what you see, smell, hear, touch, and, of course, taste. But in many eating situations your senses are useless, because the food isn't in front of you yet. In these cases, words are all you have to go on. And how can we make good choices about what and how to eat if the words used to describe our options aren't universally understood and applied?

A friend may offer you a "sandwich," but is it really a sandwich? Is a restaurant misusing language if it offers Buffalo wings that aren't from Buffalo? Does sparkling water actually sparkle—and can it even rightly be called *water*?

In this chapter I'll bring coherence to corners of the Eatscape that have been plagued by confusion. Once we've defined our terms, we'll combine words to make sentences, as we seek to further understand what effective use of language can teach us about how to eat. Finally, we'll put whole sentences together when we cover writing and poetry, and interpret some classics.

SEMANTICS AND ETYMOLOGY

Creating and consuming great works is only one part of the Eater's journey. When foods are named and described correctly, the entire Eatscape is elevated. When they are not, deliciousness is diminished.

ON THE DEFINITION OF SANDWICH

On this matter I am a strict constructionist, which means I believe we must look only at the framer's original intent to find the limits of sandwichdom. That's why I'm often called the Scalia of Sandwiches.

The Earl of Sandwich created his eponymous masterpiece because he wanted to eat substantial foods with his hands. He realized that by sandwiching meats and such between pieces of bread, he was freed up to feed his gambling addiction and/or tend to his political and military responsibilities, depending on which historian you believe.

With this origin story in mind, I've identified the two fundamental characteristics of a sandwich: First, you must be able to pick it up and eat it without utensils, and without your hands touching the fillings. Second, the fillings must be *sandwiched between* two separate, hand-ready food items.

The Earl of Sandwich's original intent is the code by which we live today, which means that bread is not a requirement. If you replace the bread with, say, fried chicken breasts or potato pancakes, you have nonetheless created a sandwich, provided it's designed to be eaten by hand. Intent is key. If it falls apart when you pick it up and the fillings get all over you, it may still be a sandwich—just a poorly made one. (For examples of a breadless sandwich see figure 2.1, plus Breadless Grilled Cheese later in this section, the Inside Out Hanukkah Miracle Sandwich on page 171 and the Stuffing Sandwich on page 159.)

A Strict Constructionist's Sandwich

This may look unorthodox, but it has the two basic criteria of a sandwich.

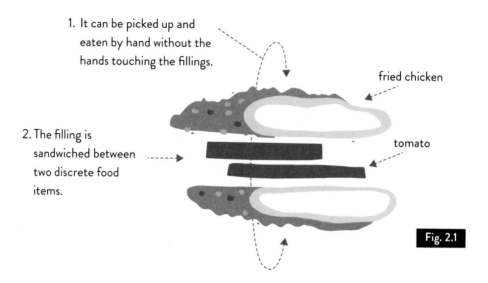

1. It can be picked up and eaten by hand without the hands touching the fillings.

fried chicken

2. The filling is sandwiched between two discrete food items.

tomato

Fig. 2.1

While the breadless sandwich fits the framer's definition, the open-faced sandwich does not. Could the Earl of Sandwich have rolled dice and/or run the Royal Navy while eating an open-faced sandwich? Not at all. Open-faced sandwiches are not sandwiches, because they require utensils and fail to sandwich anything.

For the same reason, wraps, burritos, and the like are not sandwiches. Sandwiching fillings between two things is different from wrapping them inside one.

Hot dogs, heroes, and other dishes that use a hinged bread exist in an etymological gray area, because this type of bread is technically a single structure, not two separate food items. However, these breads are capable of sandwiching, because they're essentially just rolls that have not been fully halved, and the hinge can be severed without fundamentally altering the dish. This means foods on hinged breads are sandwiches. (In fact the hinged bun is often better served severed, as discussed more in "Engineering" on page 81.)

Based on this definition, a quesadilla made with a single flour tortilla folded over itself is not a sandwich, though it may be delicious. But if you slice a flour tortilla in half and put foods in between the halves, or make a full-circle quesadilla with two tortillas, then crisp the tortilla until the structure is firm enough to be picked up, it is a sandwich.

BREADLESS GRILLED CHEESE

Eater Rex Hamaker of Virginia spent some time in the Middle East, where he acquired a passion for *haloumi* cheese. *Haloumi* is firm enough that if you coat it in oil you can put it straight on the grill and it will get crispy and charred on the outside without melting into oblivion. So put another cheese in the middle and make yourself an all-cheese grilled cheese. (Bonus tips: Season the *haloumi* with za'atar, a Middle Eastern combination of spices. For contrast, add a slice of apple in the middle.)

A REJOINDER FROM THE LIVING SANDWICHDOM CONTINGENT

NPR's Ian Chillag and Mike Danforth host the podcast *How to Do Everything*, produce the radio show *Wait Wait . . . Don't Tell Me*, and take part in the blog feature "Sandwich Monday," which is known for its especially catholic view of the sandwich oeuvre. When they came on *The Sporkful*, they argued against my strict constructionism.

"Could the Earl of Sandwich have anticipated a burrito? Could he have anticipated an open-faced sandwich?" asked Danforth. "The fault here lies in the fact that the Earl of Sandwich wouldn't have known that such things existed. He probably never even saw a tortilla."

Chillag went so far as to suggest that the simple fact that I use the term "open-faced sandwich" constitutes an acknowledgment that it is a sandwich. Let me obliterate these arguments in reverse order.

I only call it an open-faced sandwich because that's the term society has chosen. I look forward to the day that this misnomer will be replaced with a more appropriate name for the dish, such as "bread sundae."

As to the Danforth Doctrine, the beauty of the earl's vision is that it was broad enough to evolve with the times. Over the centuries we've seen the advent of many sandwiches he never could have envisioned, which still achieve his original goal. But we must draw the line somewhere, lest all foods become sandwiches.

DIFFERENTIATING CONDIMENTS, SPREADS, DIPS, SAUCES, AND INGREDIENTS

What is pesto? Tomato sauce? Cream cheese? What do you call these types of foods? In the immortal words of Supreme Court justice Potter Stewart, the difference is hard to explain, but "I know it when I see it." In other words, the difference isn't in the food itself but in how it's used. Here's a guide:

SPREAD Used as a featured component; must be spread on something else. (Example: cream cheese on a bagel or peanut butter and jelly on bread.)

CONDIMENT Used to add flavor but not as a featured component; not combined with other foods into a greater whole. (Example: mustard on a hot dog.)

DIP Served alongside and consumed with another food; viscous enough to adhere to foods inserted into it. (Example: mustard served alongside fresh-baked pretzels.) *Note:* The term "French dip sandwich" is a misnomer—the *jus* is not a *dip*, it's a condiment applied by *dunking* on a per-bite basis (see page 229).

SAUCE Adds flavor by coating or surrounding a food, but does not become one with that food; typically not eaten by itself; if bread is used to gather its remains, it is then a dip. (Example: tomato sauce served with pasta.)

INGREDIENT One component of a larger whole; not equal to or greater than the sum of the larger whole's parts. (Examples: cream cheese in cream cheese frosting; mustard in a mustard dill sauce; tomato sauce on pizza.)

ON REGIONAL FOOD NOMENCLATURE AND DEFINITION

Some of the world's most delicious dishes are strongly associated with the cities that invented them: Philly cheesesteaks, Chicago deep-dish pizza, and Cincinnati chili, to name a few. What rights should a region have to dictate the name and definition of a food that it popularizes once that food's scrumptious oils seep across borders and outsiders experiment with new preparations?

The French certify some foods as "AOC," or *appellation d'origine contrôlée*, which translates as "controlled designation of origin." Because of the AOC certification, only wine from the Bordeaux region can be called Bordeaux, and sparkling wine made in Italy is called prosecco—not champagne, which must come from the Champagne region of France.

We should adopt a similar system in the United States for foods like America's answer to Bordeaux, better known as Buffalo sauce.

I won't go so far as to say that Buffalo sauce must be made in Buffalo, but I do believe that Buffalo retains the right to determine what sauces may use that moniker. In general, the creators of a food should have the power to determine what does, and does not, constitute said food. But that doesn't mean they're the only ones who can alter it.

In fact, many regions are so immersed in their traditional ways that they become unreasonably closed to improvements. Let's look at one example of how a region that originates a food has certain authority, but also certain blind spots.

Is a pizzasteak a valid form of cheesesteak? Many of the cheesesteakeries of Philadelphia list it on their menus as a cheesesteak, which means that it is.

But when John Kerry was campaigning in Pennsylvania while running for president in 2004, he famously asked for Swiss cheese on a cheesesteak—not a traditional option and thus not a valid form of the dish.

He was lampooned, but the truth is, Swiss cheese is a perfectly delicious option for a steak sandwich. At the very least, it's superior to Cheez Whiz, a classic cheesesteak topping that epitomizes the Philly region's stubborn attachment to tradition in the face of all logic, taste, and evidence of societal progress.

It is the region's right, however, to set those parameters. In other words, Philadelphians may declare that cheesesteaks aren't made with Swiss cheese, but that does not mean that cheesesteaks aren't made *better* with Swiss cheese.

A SCREED AGAINST SIMILES, METAPHORS, ANALOGIES, AND OXYMORONS

Similes are comparisons that use "like" or "as," and they can be useful for describing a food. (What's arctic char? It's like salmon, only lighter and less oily.) But far too often, similes, metaphors, and such are used to mislead rather than illuminate.

Frozen yogurt is not like ice cream, nor is any low-fat dairy product analogous to the original, even in a metaphorical sense. Turkey burgers and veggie burgers are not "like," or "as," beef burgers. They are distinctly different foods.

I happen to be quite fond of veggie burgers. Frozen yogurt on a hot day can be very nice. These are good foods in their own right, but I have no tolerance for the saccharine entreaties of those who would offer me a turkey dog while telling me, "You really can't tell the difference."

Perhaps it's been so long since you've eaten beef, ice cream, or beef ice cream that *you* can't tell the difference. But I won't be tarred with your aspartame basting brush.

As insidious as these deceptions are, they pale in comparison to the downright oxymorons that flourish unchecked on the menus of this country. And I'm not just talking about jumbo shrimp.

Boneless ribs? THE RIB IS A BONE. To concoct a "boneless rib" is to remove the very definitional characteristic of the food. But if this type of semantic skulduggery is your bag, I hope you'll pair your oxymoronic ribs with an egg-white omelet, then wash it down with nonalcoholic beer and decaf coffee, all while listening to Sheryl Crow's cover of "Sweet Child O' Mine."

We are literally one step away from foodless food, people. Should bars begin pouring solid liquids? Perhaps for the living dead, but not for me.

ON THE PROBLEM WITH GIRL SCOUT COOKIE NAMES

Words mean something. Except when they mean nothing.

Girl Scout cookies are delectable treats that help raise money for a worthwhile organization. But while the Girl Scouts teach courage, confidence, character, and alliteration, I question whether they should do more to teach communication.

Most of the best-known Girl Scout cookies actually have two different names, depending on which of two licensed bakers is manufacturing them. (Troops decide which baker to use.)

One baker (ABC Bakers) uses literal names that give the Eater an idea of what's inside—Caramel deLites, Peanut Butter Patties, Peanut Butter Sandwiches, Shortbreads. The other (Little Brownie Bakers) uses random words that communicate nothing—Samoas, Tagalongs, Do-si-dos, Trefoils. (It's not surprising that a company called ABC shows a greater reverence for language.)

The only premiere cookie with a universal name is the Thin Mint, and while it's a fine treat, I believe the reason it's the top seller has more to do with the fact that it has the best communications strategy. Not only does it have the same name everywhere, but the name actually tells you what kind of cookie it is. That information is especially vital for a food that's only available for a few months a year. We Eaters are tired of asking, "Which ones are the Do-si-dos again?"

I don't know anything about the arrangement the Girl Scout bigwigs have with these bakers, and frankly, that's not my concern. I'm worried about the children. And that's why we must demand uniformity and clarity in Girl Scout cookie naming conventions.

Cross Section: Girl Scout Cookie Unity Cheesecake

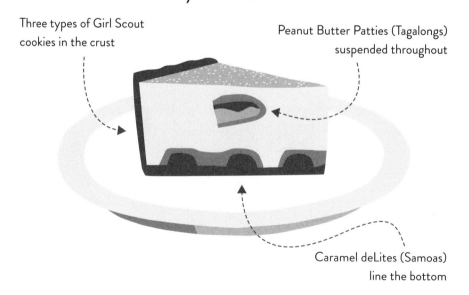

Three types of Girl Scout cookies in the crust

Peanut Butter Patties (Tagalongs) suspended throughout

Caramel deLites (Samoas) line the bottom

GIRL SCOUT COOKIE UNITY CHEESECAKE

If you unify a bunch of different Girl Scout cookies in a single cheesecake, it doesn't matter what they're called. It's Girl Scout Cookie Unity Cheesecake. And it's amazing.

My friend Emily Konn, a passionate Eater and professional pastry chef, makes the best cheesecake I've ever eaten. I asked her to create a cheesecake that uses as many types of Girl Scout cookies as possible.

CRUST

YOU WILL NEED:

8 tablespoons (½ cup) unsalted butter, melted

1 box Do-si-dos/Peanut Butter Sandwiches

1 sleeve Thin Mints

1 sleeve Trefoils/Shortbreads

1 box Samoas/Caramel deLites

INSTRUCTIONS

Adjust oven rack to middle position and heat oven to 350°. Grind all cookies except Samoas in a food processor or blender, or put in a sealable plastic bag and crush with a mallet or rolling pin into a fine meal. Mix with butter by hand and press into a 10-inch springform pan. Make sure cookies are evenly distributed along the bottom and up the sides of the pan.

Bake in oven for 15 minutes. Remove from oven and immediately layer Samoas on crust so chocolate melts slightly and helps cookies stick to the crust. Let crust cool at room temperature for 15 minutes. Transfer to refrigerator and chill completely (about 1 hour).

CHEF'S NOTE: This recipe accounts for at least two cookies being eaten out of each box during the cooking process.

FILLING

YOU WILL NEED:

24 ounces cream cheese

⅔ cup sugar

3 eggs

1 cup heavy cream

1 box Tagalongs/Peanut Butter Patties, cut into quarters

INSTRUCTIONS

Soften cream cheese until it's very mixable. You can even microwave it briefly on defrost until it starts to soften.

Place cream cheese and sugar in a bowl. Use a mixer on medium-high speed (with a paddle attachment if you have it) to beat the cream cheese and sugar together until cream cheese is smooth. Scrape sides of bowl. With mixer on medium speed, add eggs one at a time. When all eggs are added, continue to mix until thoroughly combined, about 2 minutes, scraping down the bowl twice. With mixer running on medium, gradually pour in heavy cream. Mix until just combined. (Do not overmix. If it starts to look thick like whipped cream, you overmixed.) Optional extra step: Strain it through a fine-meshed strainer to remove lumps. This gives you more margin for error if you messed up your mixing.

Pour mixture into cooled crust. Tap pan on counter to dislodge air pockets. Drop the Tagalongs evenly into the batter.

Prepare a water bath in a pan big enough to hold your cake. Crush aluminum foil together to create an S that will hold up the cake above the water level. Place cake on top

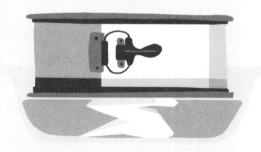

Foil and Water Bath Technique

of foil and make sure it's stable. Place in oven and pour water in pan just until it reaches the bottom of pan. If you go above the foil your crust will get soggy.

Bake for 15 minutes at 350°, then lower temperature to 250° and continue to bake for another 60 to 90 minutes or until it's firm and only the center of the cheesecake looks a little wet and wobbly (but not cracking). Let stand on rack on counter for a half hour, then refrigerate for four hours or overnight.

GIRL SCOUT PEANUT BUTTER COOKIE CENTAUR

Here's a simpler way to bring Girl Scout cookies together into something new. You'll need Peanut Butter Sandwiches (a.k.a. Do-si-dos) and Peanut Butter Patties (a.k.a. Tagalongs).

Separate top and bottom cookies of the Peanut Butter Sandwich and focus attention on half with more peanut butter on it. Lay the Peanut Butter Patty cookie upside down on that half, so Patty's peanut butter and Sandwich's peanut butter are against each other and on the bottom half of the sandwich, facing tongueward, to capitalize on the Proximity Effect (page 25 in "Physical Sciences"). Place other half of Sandwich cookie on top. Enjoy.

Peanut Butter Sandwich (peanut butter on bottom)

Peanut Butter Patty (upside down)

STEEL-CUT OATS AND THE INEVITABLE METALLURGIC ARMS RACE OF OATMEAL PREPARATION

In recent years, steel-cut oatmeal has proliferated on the menus and organic market shelves of the nation. This trend offers a valuable lesson in language.

The first time you saw the phrase "steel-cut" preceding oatmeal, did you know exactly what it meant? Did you understand precisely how it would make your oatmeal different? Probably not. But did you understand intuitively that it's considered better? Of course.

That's because they wouldn't describe it as such otherwise. It's like when you see a dish on a menu like "Happy Tree Farm Pork Shoulder." You understand intuitively that it's a good thing, and you assume that nobody from Happy Tree Farm would so much as speak to a pig unkindly.

It's true that steel-cut oatmeal is different from other oatmeals and, in many respects, better. It has more texture and holds up better over time. But do you know how other oatmeal is cut? Of course not. Nobody does.

As steel cut becomes more common and oatmeal purveyors seek to distinguish themselves, we Eaters can see a metallurgic arms race on the horizon. Soon new, more impressive-sounding methods of cleavage will arise, and steel-cut oatmeal will be relegated to prison cafeterias. Meanwhile those in the vanguard will insist upon the adamantium-cut variety. And they will know that it is better.

> **TIP** The essential difference between granola and trail mix is that granola uses a binding agent to fuse ingredients together. Trail mix—like snack mix—is a compilation, not a fusion.

WORDS THAT MAKE A FOOD SOUND SUPERIOR

Some terms get thrown around so often, we forget what they actually mean. Words like "natural" and "homemade" are popular adjectives, even though many unappealing ingredients exist in nature, and a syphilitic brothel may double as a home.

The Eater uses language more better.

Here are words and phrases you may add to seemingly ordinary creations to signal their superiority, along with examples of the usage of each:

- **DELUXE** Spaghetti and Meatballs Deluxe
- **ZESTY** Zesty Garlic Bread
- **SURPRISE** Caesar Salad Surprise
- **OF DEATH** Ice Cream Sundae of Death
- **CHEESE** Food with Cheese

MISNOMERS

Let's continue to look at the ways we describe foods, by focusing on words that describe foods incorrectly.

IT'S NOT A SCOOP, IT'S A DOME: A TORTILLA CHIP REVELATION

When scoop-shaped tortilla chips first appeared on the market, I respected the attempt at innovation. I saw that these chips allow you to get large helpings of salsa in a single bite and to build individual nachos more easily. But I also observed that they tend to break in thicker dips like guacamole and can cut the roof of your mouth. Then I had a linguistic revelation.

While researching the structural engineering of tortilla chips, I spoke with Isaac Gaetz, a licensed engineer and avowed Eater who also contributed to the engineering section of this book. He explained that a flat, triangular chip is not as sturdy as a curved one because a curved one has some properties of a dome:

> A natural, very strong shape in compression is an arch, and a 3-D arch is a dome. All the elements [in a dome] are able to take that load, work together, and just keep pushing, pushing, pushing, all the way down to wherever they're supported.

I thought about the strength of the dome, then about the scoop, then the dome, then the scoop. Dome. Scoop. Dome. Scoop. Then it hit me: *What is a scoop but an upside-down dome?*

We've been calling it by the wrong name—and holding it the wrong way—all along. When you treat a scoop like a dome, it becomes one of the strongest shapes in existence, able to withstand immense dip weight (figure 2.2)—and a prime example of how better language leads to better eating.

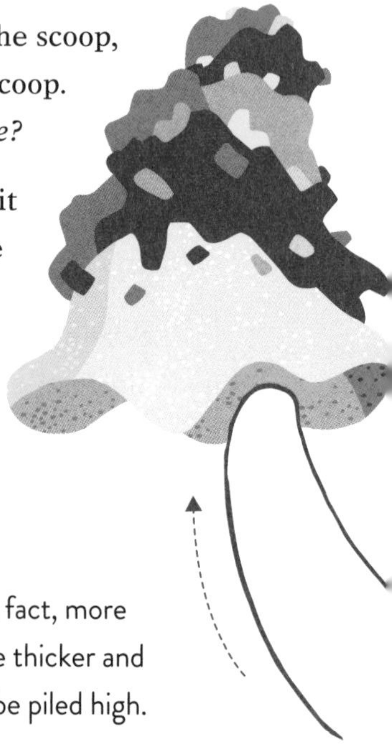

It's Not a Scoop, It's a Dome Technique for Tortilla Chip and Dip Consumption

The dome chip is able to withstand a great deal of weight. In fact, more weight is preferable. Saucier dips like salsa will slide off, while thicker and heavier options like guacamole and even cream cheese can be piled high.

Fig. 2.2

ON THE MISNOMER OF SPARKLING WATER

Water is H_2O. When you turn it into "carbonated water" by adding carbon dioxide (CO_2), carbonic acid is produced and the beverage ceases to be water. This new beverage has a different molecular structure and a different taste, violating the dictionary's dictum that water be tasteless.

However, some people argue that the added CO_2 simply seasons the water without transforming it into a different liquid. After all, they say, when you add corn syrup and food coloring to create flavored water, as certain billion-dollar companies do, isn't it still water? And why is seasoning water with carbon dioxide different from seasoning chicken with salt? It's still chicken, right?

That logic sends us down a slippery semantic slope. Water is an essential element of life and deserves to be considered differently when in its purest form. It's a basic building block of almost all foods and drinks, so if you say that when you add something to water it remains water, you'll have to consider many of the foods and all the drinks in the world to be "water."

Do we want to live in a world where chicken broth is considered flavored water? What about chicken itself, which also counts water as an integral component? Try ordering dinner at a restaurant where the only items on the menu are twenty-four variations on water.

We must draw the line somewhere, and the most logical place to draw it is at H_2O. Combine two parts hydrogen and one part oxygen and you've got water. Add anything else and it's no longer water.

It's true, however, that most of the water in the world naturally contains other minerals and compounds, so it's not simply H_2O. Therefore, I decree that if the water's chemical composition is not pure, its intentions must be. If the liquid intends to be pure water, it is. If other components (like CO_2) are added intentionally, it's no longer water.

The other problem with sparkling water is that it doesn't sparkle. It bubbles, fizzes, spritzes, and engages in other forms of onomatopoeia. But you'd need a Hollywood lighting crew to see a sparkle. Thus, "sparkling water" is a complete misnomer. Call it club soda or seltzer if you must, but I'll be calling it by the most accurate term possible: "bubbling carbonic acid." I expect sales to triple.

ARE YOU DRINKING A MILK SHAKE?

A milk shake is a drinkable dessert. It may be served with a spoon but should not require one. If you're staring at an ice-cream-and-milk-based substance, this handy chart will tell you whether it's a milkshake or some other frozen treat:

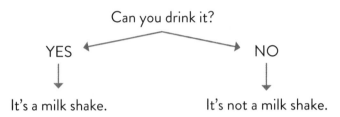

IS IT A MILK SHAKE?

Can you drink it?

YES → It's a milk shake.

NO → It's not a milk shake.

WHEN ONION RINGS ARE DONUTS: DANGERS AND OPPORTUNITIES

Great onion rings are onions first and foremost. They are not a bread-based food. Unfortunately, despite the fact that "onion" is the first word in "onion rings," far too many purveyors put the emphasis on the batter. I opposed this trend for many years.

Then I went to a place called South Carolina, where I encountered an "onion ring" with more batter than I had ever seen. A cross section was an inch in diameter. More importantly the batter was lightly sweetened, which brought out the natural sweetness of the onion. It was like a funnel cake with a thick ribbon of onion running through it. And it was amazing.

The only mistake this restaurant made was using the term "onion ring" in the first place. This was not an onion ring. It was an onion donut.

The experience caused me to revise my opinion as follows: An onion ring should have very little batter, unless it has quite a lot of batter, at which point it becomes an onion donut.

IT'S NOT WATERMELON, IT'S "WATERMELON"

Thanks to the wonders of science and the sacrifice of countless lab rats, we're all familiar with the artificial renditions of certain flavors. You've likely tried watermelon or cinnamon gum. If you've been to England you may have sampled roast beef potato chips.

These imitation flavors bear little resemblance to the originals. Most watermelon candies and gums taste pretty similar to each other, and pretty different from watermelon. As for British chips, when you attempt to reproduce the flavor of an animal's muscles, bones, tendons, and skin in the form of a powder, it tends to lose something.

Of course, your opinion of these foods is a matter of taste, so I won't tell you whether to like them. But I will tell you what to call them. If you like watermelon gum, that's fine. But it's not watermelon. It's "watermelon."

GRAMMAR AND USAGE

I feel better already. We've straightened out our terminology, lifted the shroud over our food language, and brought our goal of Perfect Deliciousness into focus.

Now that our language is clear, how will it guide our eating? Like well-chosen words, a dish should express something. Studying good sentence composition teaches us about culinary composition, and thus helps us taste the fruits and vegetables of our linguistic labors. Speaking of which, it's time we talk about salads, which offer a prime example of how the lessons of language arts apply to the gastronomic arts.

SENTENCE COMPOSITION, SALAD COMPOSITION

Good salads are like good sentences. They need balance, variation, active voice, and editing. Balance and variation in a salad come from textural and flavoral contrast. Some elements should be long and flowing, descriptive, conjuring a walk through a field of goldenrod as the sun sets behind that old oak tree where we first held hands.

Others should be sharp and short—a slap in the face.

A great salad will taste like it was crafted by the love child of David Mamet and Nicholas Sparks, with equal parts profanity and lighthouses. This is why the best salads balance contrasting ingredients like bitter and buttery lettuces, pungent cheeses and succulent fruits, sweet and acidic vegetables, crunchy nuts and soft greens.

Active voice improves salads by its mere presence. To wit: A great salad is not flavored by beets and goat cheese. Beets and goat cheese flavor a great salad. Doesn't that just *sound* more delicious?

And when editing, know that just because the word "salad" can follow everything from "tuna" to "taco," that does not mean everything belongs in one. I generally cap all salads at six total ingredients (including greens but not dressing), though many excellent salads are made of fewer. (The chicken Caesar is especially well edited.)

There's an old saying that good editing means "killing your babies," an axiom that once led me to remove baby spinach from a salad recipe. Sometimes the best decision you can make is the decision not to add something.

> **TIP** Croutons are like adverbs. They're easy to toss in, but most of them aren't very good.

SALAD LEGIBILITY AND THE KNIFE REQUIREMENT

The great writer may craft glorious prose, but if he doesn't deliver it to the reader in a form in which it can be consumed, it's all for naught.

Salad composition is no different. No matter how beautiful it looks on the plate, no matter how perfectly the flavors pair in theory, if you can't get the components together in one bite into your mouth, your salad is as useless as a great novel written in Wingdings.

The degree to which a salad can be forked and transported to the mouth with ease is referred to as its *legibility*. (See figure 2.3 on the next page for techniques to increase legibility.) Salad styles may range from the simple, classic font of the Caesar to the busy cursive of the Niçoise to the hearty block letters of the Cobb. As long as it's legible, it's acceptable. But a great debate rages within the Eatscape over where that line should be drawn, especially when a salad necessitates a knife.

Requiring a knife to eat a salad is like requiring a magnifying glass to read a book. I won't go so far as to say that you should never have to do it, but it should be rare, and there must be a very good reason. (Serving large greens whole for visual effect is not a good reason.)

Even if the logic is not all superficial, the knife requirement presents various concerns. When salad ingredients are piled on top of one another, it's hard to slice through one without at least partly slicing the food beneath. This is especially problematic when those lower strata include tender greens. If chopped too finely they'll become harder to fork, take on excess dressing, and soon resemble the dreaded chopped salad (see sidebar on page 58).

Even if you can slice some ingredients without harming others, why would you want to? Clearly, a salad is far more legible if you can just fork away until you've composed your desired bite.

The only time a knife requirement is justified is when a salad is topped with warm, freshly prepared meat or seafood. In that case, slicing the meat in advance would cause an unconscionable loss of juice and heat.

If a chef presents another strong justification, we should be open to it.

After all, the best writers often find new ways to tell stories, some of which may seem opaque at first. But remember, there's a fine line between greatness and gimmickry.

Flatter Salad Pieces Get Their Point Across

Salads communicate most effectively when their components are sliced flat instead of cut into cubes or other shapes. That's because forks are longer than they are wide and can accommodate more food when salad pieces are flatter.

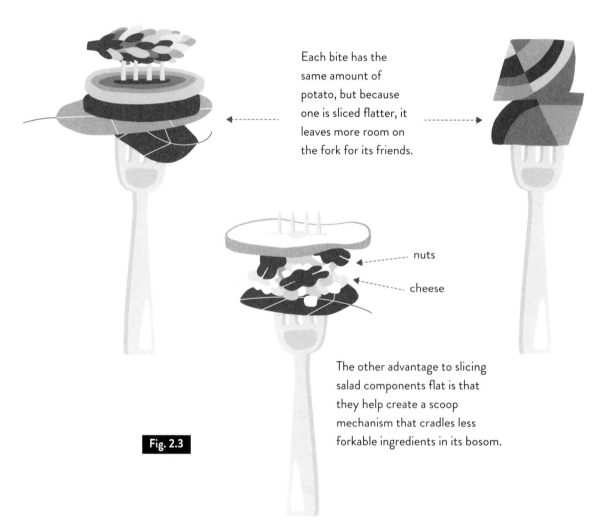

Each bite has the same amount of potato, but because one is sliced flatter, it leaves more room on the fork for its friends.

nuts

cheese

The other advantage to slicing salad components flat is that they help create a scoop mechanism that cradles less forkable ingredients in its bosom.

Fig. 2.3

INVERTING SALADS FOR BETTER BITES

Sentence inversion is when you put the verb before the subject. Not only is it a good way to vary your voice, but it also offers a useful lesson for salads. So common is the typical salad structure, you may never have thought to question it. But rarely have I inverted a salad and been disappointed with the results.

Most salads have a base of greens, with more substantial ingredients on top. This may be visually appealing, but it's practically problematic for the Eater. It means that when you drive your fork straight down through the salad, you end up with the greens on the tip of the fork, a negative for two reasons:

1. When you put that bite in your mouth, the greens will be the first thing to hit your tongue, putting the flavor emphasis on ingredients that are usually meant to be foundational, not featured. (This is another example of the Proximity Effect discussed in the previous chapter.)

2. When greens are at the fork's tip, they're more likely to fall off before they make it to your mouth.

The inverted salad addresses both of these issues, and also makes it easier to form a functional scooping mechanism (see figure 2.3).

If you're served a salad that's built in the traditional way (uninverted), don't despair. One of the great things about salads is that you have the power to create so many different experiences with the same basic ingredients. You're not beholden to the layering set forth by the chef. Put thought into how you assemble each bite. Fork pieces in ascending order of flavor priority, so the taste you want to accentuate ends up at the tip and thus lands directly on your tongue.

DON'T JUST EDIT CHOPPED SALADS—DELETE THEM

Chopped salads are abominations that have been foisted upon us as if they're both gourmet and convenient, when in fact they're neither. These monstrosities are completely unforkable and can only be eaten with any degree of comfort with a spoon. But that only solves part of the problem. The high surface-area-to-volume ratio of chopped greens also causes them to absorb dressing at a destructive pace, lending them a mealy consistency.

It's true that when done perfectly, a chopped salad can be enjoyable. But 99 percent of the time, they're an excuse to use cheap greens in volume, because you can make a huge batch at once, slather it in dressing, and parcel it out like gruel.

Do not eat chopped salads.

CHERRY TOMATOES: TO HALVE OR TO WHOLE?

Cherry tomatoes are delicious when they're in season, but when served whole they present a serious challenge to the tenets of good salad composition. Like an especially long dependent clause that just keeps going on and on and makes a sentence feel grammatically incorrect even when it technically isn't, a whole cherry tomato in a salad is simply unwieldy.

When an intact cherry tomato is pierced by the teeth, it releases a deluge. This means a whole one must be fully inserted into the mouth before biting down, to avoid splatter. As a result, these devils dominate any bite in which they're included.

Furthermore, forking a whole cherry tomato is quite difficult, because its round shape and taut skin resist piercing. The best way to enjoy great cherry tomatoes is whole, à la carte, and with your hands. But if you wish to include them in a salad, halve them before you have them.

What about grape tomatoes, you ask? Their smaller size alleviates some of my concerns, but I've never had a great one. The grape tomato's shape seems to disturb the fruit's natural interior balance between meat and pulp, to the point that I believe they'd be more accurately called tomato grapes.

CRISPY SALAD POPPERS

Here's a moderately healthy way to eat junk food. Start with a big bowl or bag of your favorite greens, without dressing, and a small bag of your favorite flat junk food, like potato chips. Smaller chips work better so you can eat the whole thing in one bite and cut down on chip splintering. Note that you can put a chip on top also, to make a sandwich, but that may lead to cutting the roof of your mouth, the condition known as Cap'n Crunch's Complaint. (You could wrap the chip in greens, which reduces splintering and protects the roof of your mouth, but then the chip isn't directly on your tongue.)

Junk food goes on the bottom, so it's against your tongue.

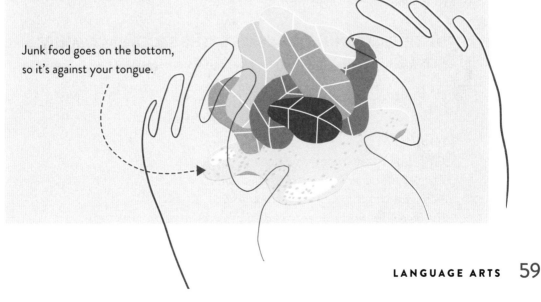

POETRY AND COMPOSITION

Like food, language can lift our spirits to great heights in a way that's often hard to describe. It seems to work by magic, as if an idea or feeling has been conjured from thin air. Eaters, it's time for us to soar.

INTERPRETING POETRY

In my freshman year of high school, my teacher Mr. Dunn read us a poem. I don't remember exactly what it was about, maybe flowers or a walk in the park. He then went back through the poem, line by line, pointing out specific word choices and connections, to explain that it was actually about much more—love and loss, hope and fear, etc.

While I don't remember the poem or Mr. Dunn's interpretation clearly, I have a vivid recollection of the reaction in the classroom. We looked at each other and snickered, thinking that this guy pulled this whole other meaning out of nowhere. To our fourteen-year-old minds, the idea that the word "flower" might not be referring to a flower was ridiculous.

Over time, of course, we learned that a lot of great works of poetry and literature use symbolism to address deeper themes, and I came to love interpreting poetry. Now I hope to pass that passion on to you, as we dissect one of my all-time favorite works.

THE TRUE MEANING OF YEATS'S "THE SECOND COMING"

Let's begin by reading this classic poem in its entirety.

THE SECOND COMING
by William Butler Yeats

Turning and turning in the widening gyre
The falcon cannot hear the falconer;
Things fall apart; the centre cannot hold;
Mere anarchy is loosed upon the world,
The blood-dimmed tide is loosed, and everywhere

The ceremony of innocence is drowned;
The best lack all conviction, while the worst
Are full of passionate intensity.

Surely some revelation is at hand;
Surely the Second Coming is at hand.
The Second Coming! Hardly are those words out
When a vast image out of Spiritus Mundi
Troubles my sight: somewhere in the sands of the desert
A shape with lion body and the head of a man,
A gaze blank and pitiless as the sun,
Is moving its slow thighs, while all about it
Reel shadows of the indignant desert birds.

The darkness drops again; but now I know
That twenty centuries of stony sleep
Were vexed to nightmare by a rocking cradle,
And what rough beast, its hour come round at last,
Slouches towards Bethlehem to be born?

Wow. Are you guys sweating too? I know that's a lot to process, so let's go back through the poem line by line so we can break it down.

Turning and turning in the widening gyre
The falcon cannot hear the falconer;
Things fall apart; the centre cannot hold; . . .

A gyre is a spiral or vortex, like a tornado. We know that the gyre is widening and that it's loud and chaotic, because the falcon can't hear the falconer. We also know that things are falling apart, because Yeats writes, "Things fall apart."

But what is he really referring to? It sounds like a big storm, but that's too obvious. I believe he's using symbolism. So let's look across the Eatscape for clues. What is a situation that involves a sort of vortex, where Eaters may be pulled in different directions against our will, where noise and chaos threaten our pursuit of deliciousness?

Of course, it is an all-you-can-eat buffet. And this gyre of a buffet is very crowded, so getting to the foods we want is quite difficult.

Now that we've figured that out, this poem starts to make more sense. The falcon is the Eater—a fierce bird of prey—and the falconer represents our training. The buffet Yeats describes is so intense, we are unable to hear the voices of our instructors reminding us of proper buffet strategy. (For a detailed discussion of buffet strategy, see page 127 in the "Business & Economics" chapter.)

But why are things falling apart? Why can't the centre hold? Let's keep reading . . .

> *Mere anarchy is loosed upon the world,*
> *The blood-dimmed tide is loosed, and everywhere*
> *The ceremony of innocence is drowned;*
> *The best lack all conviction, while the worst*
> *Are full of passionate intensity.*

Clearly, this is not your ideal buffet. A blood-dimmed tide has been released, and this blood must be a reference to meat, the only logical source of blood in this context.

Now, I know what you're thinking: "A blood-dimmed tide sounds delicious. What's the problem?"

I had the same reaction the first time I read this poem. But note the use of the word "loosed" here. A blood-dimmed tide has been loosed, or released, meaning it's been allowed to *escape*. We should be drowning in meat, but instead we've learned a cruel lesson about buffets, and it's our "innocence" that's been drowned.

And if an absence of meat has caused us to lose our innocence, that can only mean one thing. *THE ALL-YOU-CAN-EAT BUFFET IS OUT OF PRIME RIB!* People are going crazy. Even the best lack all conviction.

This would be a pretty bleak poem if it ended there. But it's not called "The Second Coming" for nothing. Just after Yeats brings us down to rock bottom, he gives us reason to hope . . .

> *Surely some revelation is at hand;*
> *Surely the Second Coming is at hand.*

Yes, surely!

> *The Second Coming! Hardly are those words out*
> *When a vast image out of* Spiritus Mundi
> *Troubles my sight: somewhere in the sands of the desert*
> *A shape with lion body and the head of a man,*
> *A gaze blank and pitiless as the sun,*
> *Is moving its slow thighs, while all about it*
> *Reel shadows of the indignant desert birds.*

Hallelujah, we are saved! Eaters are so happy to see a second helping of beef that we're practically hallucinating. The replenishments seem like an image from a spirit world, springing forth from a desert oasis. Yeats refers to the cut of meat as a lion, the king of the jungle, to illustrate its primacy at the buffet. And he describes the man carrying it out from the kitchen as "blank and piti-less . . . moving [his] slow thighs," because his cheap boss was really hoping they wouldn't have to cut into another prime rib before the dinner seating.

Meanwhile, roast chicken breasts look on indignantly, wondering why no-body cares when the buffet runs out of *them*. But Yeats knows it's because they're dry, which is why he calls them "desert birds."

Now let's look at the conclusion . . .

> *The darkness drops again; but now I know*
> *That twenty centuries of stony sleep*
> *Were vexed to nightmare by a rocking cradle,*
> *And what rough beast, its hour come round at last,*
> *Slouches towards Bethlehem to be born?*

As night falls, Yeats acknowledges the trauma of dining at an understocked all-you-can-eat buffet. He points the blame at a "rocking cradle," which I think is a reference not to a baby's cradle, but rather to the framework on which a boat rests when it's repaired. In other words, the evil Buffet Master, whom we'll discuss in more detail in "Business and Economics," rocked the buffet boat by failing to have more prime rib ready in a timely fashion.

As the poem ends, Yeats considers the animals ("rough beasts") being born now that will make their way to our buffet serving stations, hopefully eliminating the type of shortfall experienced today. By referencing Bethlehem, the birthplace of Jesus, he calls these animals saviors, paying tribute to their sacrifice and ultimate demise for our benefit.

What a masterpiece!

THE WRITER'S TOOLS

When composing a meal, it's important to understand the techniques you may employ:

- **CHARACTER AND MOTIVATION** The characters are the people at the meal. Their motivation is to eat a lot of delicious food.

- **EXPOSITION** A menu.

- **PLOT AND CONFLICT** Meals should have a beginning, middle, and end, with a tension between flavors and textures that's resolved in the conclusion. People don't like meals that lack a clear resolution, because nobody wants to go to bed hungry.

- **HYPERBOLE** A very large meal.

- **FORESHADOWING** When you say, "I think it's time for the next course."

- **IRONY** When a dish you cook comes out poorly.

- **SUSPENSION OF DISBELIEF** When you're served something ironic at someone else's house and, until you get home, you suspend your disbelief.

GEORGE ORWELL:
WRONG ON 1984, WRONG ON TEA

Interpreting poetry and fiction is an important skill, but you also need to be able to analyze and critique an argument. You might know George Orwell as the author who failed to predict anything that would happen in 1984. But I know him as the author who failed to argue persuasively for the superiority of loose tea over bagged tea.

In his famous 1946 essay "A Nice Cup of Tea," Orwell lays out eleven rules for his beloved beverage. To be fair, he's not all wrong, and in fact, he makes some good points: You should make tea in small batches, no larger than a teapot. Use boiling water. Make it strong, and without sugar.

But he blunders when he insists that only loose tea, floating directly in the water, will do, saying that if it's in a bag or net of any kind, it never steeps properly. Clearly he's guilty of the same thoughtcrime that has enticed fancy-tea drinkers for decades.

High-end teas tend to be sold loose, and lower-quality ones tend to be sold in tea bags. As a result, many tea drinkers mistake correlation for causation and assume the delivery system is the cause of the difference in quality, when in fact it's simply the difference in the teas themselves.

Orwell then offers an absurd method for steeping that lays bare one of the shortcomings of loose tea. He says that after combining the water and leaves, you should stir the tea and "give the pot a good shake."

Really, George? You're going to shake the whole teapot full of boiling water? That's a rather vulgar approach for a guy who spells "flavor" with a "u." And all it does is achieve the exact same result as bobbing a tea bag up and down in the water, while increasing the chance you'll burn your face off.

Tea bags offer additional advantages. The leaves are consistently portioned, so you can always steep to your liking. And there are no dishes to wash, as there are with the metal devices that hold loose leaves. But I guess that's just too much logic for Mr. Orwell. Sounds like someone needs a visit from the Thought Police.

HAIKU

Haiku is a beautiful style of Japanese poetry with strict rules for its composition. It must be three lines consisting of five, seven, and five syllables respectively. When this structure is followed, almost any combination of words soars. For instance:

My stomach exults
in perfect deliciousness
I dream I eat dreams

Try writing your own haiku about your favorite bite!

HOMEWORK

While you were reading this chapter, I'll bet you realized there are at least a few terms you've been using incorrectly, and that poor sentence structure may have had a negative effect on your salads. I also suspect you never knew the true meaning of "The Second Coming."

Now that we're all on the same page, we'll be able to communicate more effectively, avoid confusion in the Eatscape, and appreciate great works as a community.

For your assignment, read the classic Shakespeare sonnet below and interpret it.

SONNET 18

by William Shakespeare

Shall I compare thee to a summer's day?
Thou art more lovely and more temperate:
Rough winds do shake the darling buds of May,
And summer's lease hath all too short a date:
Sometime too hot the eye of heaven shines,
And often is his gold complexion dimm'd;
And every fair from fair sometime declines,
By chance, or nature's changing course, untrimm'd;
But thy eternal summer shall not fade,
Nor lose possession of that fair thou ow'st;
Nor shall Death brag thou wander'st in his shade,
When in eternal lines to time thou grow'st;
So long as men can breathe, or eyes can see,
So long lives this, and this gives life to thee.

Hint: It has something to do with General Tso's Chicken.

Submit your homework to me at dan@sporkful.com.

3

ENGINEERING
Construction as Cookery

- Newton's laws of motion and elementary sandwich construction

- The Sliced Cucumber Conundrum

- Bread-based sandwich foundations

- The Semolina Fulcrum

- Wrap techniques: the High Horse, the Snail's Cochlea, El Mixtec

- Improving pancakes and bacon with the Porklift

- Architectural masterpieces Fallingsandwich and the Geodesic Breakfast Dome

- The construction and consumption of waffles and omelets

Whoever said Americans don't make things has obviously never eaten a sandwich. Eaters across the country assemble edible edifices every day. Some build for height, reaching for the heavens of Perfect Deliciousness. Others push the boundaries of structural integrity, seeking to tame the most uncooperative ingredients. Whether your vision is daring or mundane, it can be improved by the application of basic engineering principles.

This chapter is your bridge from pie-in-the-sky ideas to pie-on-your-plate realities. We'll build a bacon lattice structure called the Porklift to elevate pancakes and reduce soggage, and erect a tribute to architect Frank Lloyd Wright in the form of Fallingsandwich, a Buffalo chicken sandwich cantilevered over a waterfall of blue cheese dressing.

But we'll begin with more everyday examples of how engineering affects eating. If you want bite consistency in your wrap, for instance, you must build it into your design for the fillings, the wrap construction, or both. Failure to do so is like erecting a skyscraper without elevators. If you want a sandwich with a crispy exterior and juicy interior, you must consider bread foundation, load-bearing capacity, and moisture movement. Failure to do so is like commissioning a palace on quicksand.

Not all structures are engineered equal—some are proud temples, while others serve as cautionary tales of the importance of fire escapes. Build your foods well, and the pleasure they provide will live on for generations.

ISAAC NEWTON AND SANDWICH MOTION

To construct a great sandwich, you must consider how the ingredients come together not only on the palate, but also on the plate—and in your mouth, as your fearsome jaws work to lay low your mouthwatering monument.

A sandwich may look static, but because it's subject to both gravity and repeated bite force, its components are in near-constant motion. Effective sandwich engineering accounts for this motion by looking to Newton's laws. These blessed rules form much of the basis of classical mechanics, which remains to this day one of the most delicious branches of mechanics.

TIP Certain words evoke a negative visceral reaction whenever people hear them. Linguists call it word aversion, and apparently, "moist" is high on the list. If you're one of the people who can't handle that word, I hope you can at least stomach "moisture," because we're going to talk about it a lot in this chapter.

FIRST LAW: AN OBJECT IN MOTION
(or, Regulating Sandwich Moisture)

Newton's First Law states that an object in motion will stay in motion and an object at rest will stay at rest, unless unbalanced forces act upon them.

Those unbalanced forces are you. And if you act upon moisture correctly, your sandwich will stay in motion until it comes to rest in your mouth.

Moisture in a sandwich can come in the form of condiments, spreads, sauces, cheesy oils, and meat juices, to name a few. Regardless of its source, it is the great double-edged sword of sandwichdom. You need it, not only because overly dry foods are unpalatable, but also because it's a binding agent vital to structural integrity. However, too much moisture eats away at your bread foundation and can reduce your repast to ruins.

Let's cover a few principles:

- **GRAVITY PULLS MOISTURE EARTHWARD.** Pay special attention to the bread on the bottom of the sandwich, which may be more susceptible to soggage-induced mechanical compromise as fluid flows downward. Thin liquids like gravy and meat juice may remain in motion until the bottom bread brings them to rest. In a perfect world you'd add these liquids at the last minute, or dunk the sandwich in them on a per-bite basis so the bread doesn't have time to deteriorate on its way to your mouth. But if you're on the go, note that the domed top half of some rolls is thicker and thus more absorbent than the flat bottom. If you're taking a sandwich on a roll to go, store it upside down so moisture moves toward the bread more capable of accommodating it. (See figure 3.1, "Good for the *Jus*.")

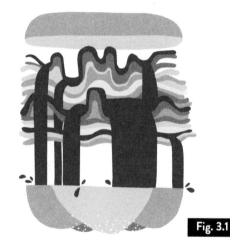

Good for the Jus

When on the go, store moist sandwiches on rolls upside down. Gravity pulls juice down toward the "top" bun, which is thicker and thus more able to withstand the flood.

Fig. 3.1

- **THERE'S MORE THAN ONE WAY TO BRING MOISTURE IN MOTION TO REST.** Containing sandwich moisture is about not just how thick the bread is, but also how dense, crusty, or toasted it is. For instance, olive oil can be delightful on a sandwich, but pouring it on even the thickest supermarket white bread is like asking a thatched hut to weather a monsoon. Olive oil requires crusty or toasted bread to contain it.

- **SEND REINFORCEMENTS.** Don't ask your bread to withstand moisture by itself. Add melted cheese, which is a great sealant, or use thin layers of greens to line the bread (see figure 3.2, "A Silver Lining of Greens"). This may increase greens soggage, but if you use whole dry leaves, the moisture will be a welcome addition. (Never use shredded lettuce, which has a high surface-area-to-volume ratio. Salt and fat attack it from all sides so it breaks down quickly, releasing water and turning to mush.)

A Silver Lining of Greens

Place greens strategically throughout the sandwich instead of concentrating them at the top. This provides absorbency and friction between layers and protection for the bread.

Fig. 3.2

SECOND LAW: FORCE = MASS X ACCELERATION
(or, Preventing Sandwich Slippage)

How do you build a sandwich so that the force of your bite does not cause the mass of the sandwich to accelerate out the back? Companies that make stain-resistant pants hope you never find out. But I'll tell you.

Two concerns are at play here: the hardness of the sandwich's exterior and the slipperiness of its interior. The sturdier the bread, the more bite force you need to pierce it, which means more pressure applied to the interior, which means a greater chance that slippery or loose fillings will accelerate lapward. As exterior crustiness increases, so must interior ingredient friction. Conversely, as fillings become more slippery, bread must become softer, so less bite force is required. The point at which this relationship becomes structurally sound is the Semolina Fulcrum (see figure 3.3).

The Semolina Fulcrum

This specifies the point at which the relationship between sandwich bread hardness and filling stability becomes structurally sustainable.

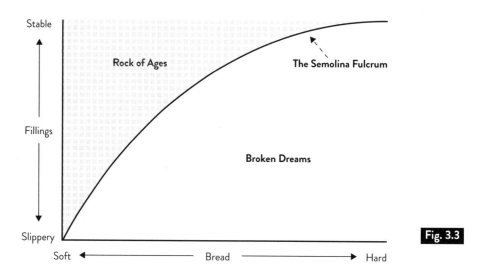

Fig. 3.3

We'll talk more about bread stiffness in the next section of this chapter. For now, let's focus on a classic problem of interior sandwich stability—the Sliced Cucumber Conundrum and its corollaries, the Sliced Tomato Botheration and the Sliced Avocado Gordian Knot (see figure 3.4).

The Sliced Cucumber Conundrum

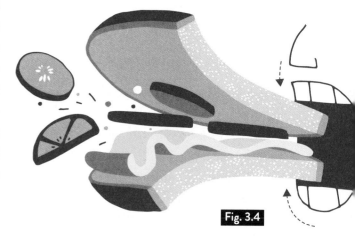

Fig. 3.4

Sliced tomatoes, avocados, and cucumbers are wonderful sandwich ingredients, although the physics of using all three in the same sandwich successfully are almost unthinkable. (I've done it. But you may not be ready yet.)

If you plan to use two of the three in a sandwich, you must never place them against each other. Doing so virtually guarantees slippage, slideage, and worse.

The solution is friction.

Place a single layer of dry greens between the tomatoes and cucumbers. (See also figure 3.2, "A Silver Lining of Greens," as this is essentially the same technique used for a different purpose.) I recommend baby spinach, arugula, or mesclun mix, perhaps with a frisson of frisée—a spiky plant with a high surface-area-to-volume ratio (SATVOR), meaning it will provide lots of friction.

Romaine and iceberg lettuce work well if you're using whole leaves, but if you end up with a spine in there, you're in trouble. Even moderate force against their smooth, hard surface can launch a tomato across the room.

> **TIP** The higher SATVOR a sandwich filling has, the more it will grip whatever it touches. For instance, a chicken breast with rough grill-mark indentations will be more stable than one with a smooth, roasted surface.

The key here is to keep the layers of greens thin—one or two Leaf Thickness Units (LTUs), where one LTU = one leaf. This way you can have a structurally sound creation without making a sandwich more fit for a rabbit.

THIRD LAW: FOR EVERY ACTION
(or, Maintaining Sandwich Equilibrium)

Newton's Third Law of Motion states that for every action, there is an equal and opposite reaction. Underlying this statement is an almost philosophical notion of a natural balance maintained throughout the sandwich universe. In Eatscape terms, this balance is bite consistency.

Bite consistency must be the goal in all sandwich construction, because it's inherent in the form. Why else would we build a food vertically in layers, then consume it in horizontal bites, if not to get equal parts of each layer in each bite? The only acceptable source of bite variety in a sandwich is the natural variation between sandwich perimeter and center.

If you seek bite variety, many other foods can provide it as a natural by-product. A steak offers fat and lean. A salad lets you fork a wide range of combinations. But the sandwich form is designed for bite consistency.

Great engineers seek new and innovative ways to attain, and improve upon, this sandwich balance. Of course, the typical way to layer ingredients in a sandwich is: meat, cheese, veggies, greens. But have you ever stopped to consider why?

People naturally begin with the focal filling, for which the sandwich is often named, then add the other fillings in descending order of significance. And putting heavier ingredients on the bottom does lower the sandwich's center of gravity and improve stability.

But this classic technique is also fraught. The separate strata don't truly come together into one until they're halfway down your throat. Until then you often have bites where the top third is veggies and the bottom two-thirds are meat. With care, however, an Eater can create a cohesive structure with true bite consistency that's greater than the sum of its ingredients.

Enter my Symmetrical Sandwich (see recipe below), inspired by Newton's lesser-known Fourth Law of Motion, "For every half sandwich there is an equal and opposite half sandwich."

SYMMETRICAL SANDWICH

This sandwich provides a visually striking cross section, as well as a level of balance and bite consistency previously thought impossible.

From bottom to top, layer your sandwich as follows:

Russian dressing
1 slice roast beef
1 slice American cheese
1 slice roast beef
1 slice tomato
2 layers arugula
1 slice tomato
1 slice roast beef
1 slice American cheese
1 slice roast beef
Russian dressing

Fig. 3.5

Note: The bread is buttered and griddle-grilled, but that has nothing to do with being symmetrical; it just tastes better.

These particular ingredients are just suggestions. Try symmetrically layering your favorite sandwich and experience the joy of equilibrium.

PRINCIPLES OF SANDWICH ENGINEERING AND ARCHITECTURE

Newton's laws of motion form a foundation for the study of sandwich engineering, but when it comes to forming foundations for sandwiches themselves, you want to use actual food. What kind, and with what properties? And once you know you're on solid ground, what can be done to add form to function?

ON LOAD BEARING IN SANDWICH CONSTRUCTION

No sandwich can be structurally sound without a good foundation and adequate support, and the bread is primarily responsible for providing these elements. Two principles are most vital to sandwich bread: strength and stiffness.

Strength is the bread's ability to support the sandwich mass, especially in tension from compressive loading (biting) and cyclic loading (chewing). Stiffness is a measure of structural rigidity, determined by the degree to which a sandwich bread has been toasted or griddle-grilled before sandwichization. While more stiffness does mean more support, if sandwich bread is too stiff, the additional bite force required may cause buckling and the collapse of the structure.

All this talk of strength and stiffness is not to suggest that vulnerable breads should never be sandwichized. Thor Heyerdahl rowed across the Pacific on a raft made of twigs. You just have to know what you're doing. And it helps if your name is Thor.

This table should help you pick your foundation:

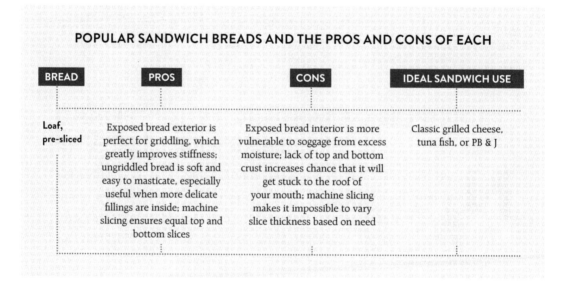

POPULAR SANDWICH BREADS AND THE PROS AND CONS OF EACH

BREAD	PROS	CONS	IDEAL SANDWICH USE
Loaf, pre-sliced	Exposed bread exterior is perfect for griddling, which greatly improves stiffness; ungriddled bread is soft and easy to masticate, especially useful when more delicate fillings are inside; machine slicing ensures equal top and bottom slices	Exposed bread interior is more vulnerable to soggage from excess moisture; lack of top and bottom crust increases chance that it will get stuck to the roof of your mouth; machine slicing makes it impossible to vary slice thickness based on need	Classic grilled cheese, tuna fish, or PB & J

Loaf, hand-sliced	Same as above, except that hand slicing offers ability to vary thickness on a per-sandwich basis	Same as above, except that hand slicing makes it harder to achieve consistent thickness, not only between top and bottom slices but also within each individual slice—a slanted interior is a risk	Steak sandwich on buttered, grilled bread, to be dipped into gravy or *jus* on a per-bite basis
Roll (round or oblong)	Thick enough to handle moderate moisture; exterior crust increases strength; domed top can often accommodate otherwise-errant fillings like sun-dried tomatoes or broccoli	Exterior is not greatly improved by toasting or grilling, and it may be hard to heat it through to the center without drying outer layers	Broccoli, ricotta salata, pickled lychee, and peanuts, as served at *Sporkful* guest Tyler Kord's No. 7 Sub in New York.
Bagel	Possesses a particular type of bulk that pairs well with subtle additions like melted cheese, as well as spreads like cream cheese	Often too thick, dense, and hard, requiring so much bite force that fillings are pushed out the back; see Bagel Trifurcation Technique on page 199 for a tip	Cream cheese and lox (duh); melted cheese with warm deli meats, provided cheese is sliced thin—too much cheesy oil + bite force required by a bagel = trouble
Baguette	Hard crust is strong and can tolerate high levels of moisture	Hard crust requires extreme bite force and can cut the roof of one's mouth, resulting in the condition we call Cap'n Crunch's Complaint	Warm prosciutto, mozzarella, and arugula with olive oil
English muffin	Unique flavor profile; nooks and crannies may act as safe harbors for excess condimentation	Small and thin, meaning sandwich size must be minimal and moisture must be curtailed	Toasted, buttered, and filled with good ham
Croissant	As a sandwich bread—none	Too soft and delicate for sandwichization; the mere act of slicing it in half destroys the croissant's fragile beauty	Because don't.
Donut	Offers opportunity for sweet/ savory combos and dessert sandwiches; light toasting improves texture and support	Sweetness of many donuts is overpowering; large center hole presents design challenges	Balance a yeast donut's soft texture and sweet flavor with salty crunch. Yes, that means bacon.

SANDWICH ARCHITECTURE AND FORM VS. FUNCTION

Architects and engineers both think they have the more important job, but in truth, they each have their place. Without architects buildings would be unattractive, and without engineers they'd all fall down.

In other words, engineers are more important, because you'd rather be standing in an ugly building than crushed beneath a pretty one.

But the architecture is what people see, and it's true that beautiful structures make the world a better place. So it's always the Eater's goal to combine form and function, even though that's easier said than done. Here are examples of sandwiches that fall short:

- **FUNCTION WITHOUT FORM** Consider turkey and mayo on white bread. It may be perfectly delicious and structurally impeccable, but it's so white and pale that even when fresh, it looks like it's been sitting out. Add some color. If you oppose vegetables, at least use whole wheat bread for a bit of contrast.

- **FORM WITHOUT FUNCTION** Consider a sandwich with many jagged components protruding beyond the bread perimeter in all directions. This rough-hewn style is often used to emphasize a handmade provenance or hearty portion, but it only works if you can trim the protrusions with precise nibbling. Too often, that nibbling disrupts structural integrity, because it pulls a filling out entirely, or because the mouth can't get close enough to the bread to take a stable bite.

When form and function are at odds, function must always triumph. But the great sandwiches have both style and substance.

So as you begin construction, don't just break ground—break new ground. Strive to incorporate color contrast, textural variety, and a ratio between height and footprint that accentuates your creation's aesthetic attributes as well as its deliciousness.

TO SLICE OR NOT TO SLICE

Slicing a sandwich in half exposes its interior, which improves visual presentation and offers greater choice in bites. But if that slicing puts sandwich integrity at risk, it's not worth it.

When deciding whether to slice, consider the following:

- **WHAT IS THE SANDWICH'S FOOTPRINT IN RELATION TO ITS HEIGHT?** A flat, wide structure is more stable than a tall, narrow one. That's why some restaurants use toothpicks to stabilize triple-decker obelisk sandwiches.

- **DOES THE SANDWICH CONTAIN MELTED CHEESE?** If so, how much? Moderate levels generally act as a binding agent, while high levels may have the opposite effect.

- **HOW MUCH MOISTURE IS PRESENT?** Dry sandwiches have little to hold them together, and wet sandwiches are just asking to fall apart. If you're in the sweet spot, sliceability increases.

WHY ROLLS SHOULD COME UNHINGED

Many sandwiches, subs, and heroes are served on long rolls that are sliced lengthwise without being fully severed in two. The rolls are then rotated into a V position and filled from the bottom of the V to the top. This method leads to bad sandwiches and bad eating experiences.

It's both difficult and uncomfortable to try to eat a sandwich when it's inserted into your mouth in the V formation. The only bites that contain all the ingredients are the ones in the center of the V, which have no top bread. But when you rotate the roll back to its natural position for eating, some fillings are likely to fall out, and the ones that remain are now arranged from left to right instead of bottom to top, so it's even harder to get them all in the same bite (see figure 3.6).

A Hinged Roll That's Filled Leaves the Eater Unfulfilled

In this construction, there's no vertical axis offering bite consistency and correct proportions of all the ingredients, except in the center of the V, where

there is no top bread. The easiest solution is to deconstruct and rebuild the sandwich by hand. Sever the hinge and layer ingredients as you like. Cap it with the top and enjoy.

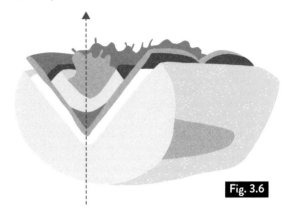

Fig. 3.6

WHEN LIQUID CONTAINMENT HINGES ON HINGES

The only time a hinged roll or bun is preferable is when there is a substantial liquid element, such as with a chili cheese dog or meatball sub, so that the hinged bun's similarities to a bowl become advantageous.

Of course, few hinges can stand up to much liquid, which is why I recommend reinforcing them with melted cheese (figure 3.7).

Buttressing the Hinge with Cheese

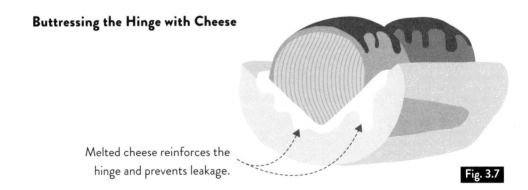

Melted cheese reinforces the hinge and prevents leakage.

Fig. 3.7

SANDWICHIZING ALTERNATIVE MATERIALS

This chapter has focused mostly on traditional methods—that is to say, sandwiches made with bread. As explained on page 37 in "Language Arts," however, bread is not a prerequisite for a sandwich. Furthermore, principles of sandwich engineering can improve all sorts of foods, even when those foods are not actually sandwiches.

DON'T BAKE BETTER CUPCAKES—BUILD THEM

There are a lot of mediocre cupcakes out there. Cupcakes are primarily frosting delivery systems, but the cupcake community often forgets about the cake entirely, leaving that portion bland and dry.

The simplest way to improve a cupcake is to eat it upside down, so the frosting is placed directly on your tongue, thus capitalizing on the Proximity Effect. Another way to bring the frosting to the fore is through ratio management, by removing the bottom half of the cupcake and offering it to an unsuspecting guest, small child, or pet. Combine both techniques and you're essentially eating frosting with a cake chaser, about as close to eating pure frosting as you're likely to get in mixed company.

But if you're looking for a more subtle and elegant way to focus on the frosting, try Cupcake Sandwichization. Remove the bottom half of the cupcake and place it on top. This brings frosting closer to your tongue and also makes the cupcake easier to hold and bite.

SANDWICHESQUE: A GRAY AREA

My dear pupils, you're far enough along in this course to move beyond black and white, bread and filling. The foods I'm about to share do not fit the strict-constructionist definition of a sandwich. But they make such good use of sandwich principles that we can call them *sandwichesque.*

THE FIXINS

The lettuce, tomato, onion, and pickle you typically get with a burger are called the Fixins, and they are often problematic. The veggies are sliced so thick that they threaten to overpower the burger, and served so cold that they chill both meat and cheese. Unless great care has been put into the fixing of the Fixins, they're better enjoyed on their own, where their strong, contrasting flavors are assets.

Use a Lettuce Glove Technique to pick up and enclose the tomato, onion, and pickle. Dip into the condiment(s) of your choice on a per-bite basis and add salt and pepper to taste. My recommendation: I love how ketchup's sweetness complements a pickle's zing and how mayo's cool soothes an onion's bite.

FALLINGSANDWICH

Famed architect Frank Lloyd Wright built a house on a waterfall that he called Fallingwater. This dish is my tribute. It's a tower of buffalo chicken sandwiches, cantilevered, over a waterfall of bleu cheese dressing and bacon bits. The dressing flows down a wedge of iceberg lettuce, the tower is anchored by bamboo skewers, and celery stalks provide horizontal support for the cantilever. And yes, I did this in real life.

What great work of architecture would you like to honor in sandwich form?

WRAPS: A SURVEY

Let's move on from sandwiches, as there are many other foods in the Eatscape that are also improved through better engineering. Wraps offer a good segue, because while they are not sandwiches themselves (as explained on page 38 of "Language Arts"), they are similar.

WRAPS VS. SANDWICHES

To understand the promise and the pitfalls of wraps, it helps to continue the comparison to sandwiches. Both employ often starchy exteriors to allow fillings to be hand-delivered to the mouth. So how is a wrap superior to a sandwich?

- Because it fully encases the fillings, a wrap can accommodate small pieces of food in a way that a sandwich can't. Foods like rice and beans would never stay in a sandwich, but they'll stay in a wrap.

- It offers a much lower ratio of bready starch to interior, which puts the focus on the fillings. To achieve the same ratio using sandwich bread would require impossibly thin slices.

- The wrap's circularity offers the Eater three hundred sixty different angles from which to attack.

How is a wrap inferior to a sandwich?

- Because they're so thin, wraps can handle little moisture or condimentation, and tear easily when exposed to sharp ingredients like tortilla chips or even lettuce spines.

- Because the physical act of wrapping tends to force fillings into clusters, it's harder—though certainly not impossible—to achieve bite consistency.

- In a warm wrap, crunchy fillings may come under fire. The steamy, sealed interior ecosystem creates almost instant condensation—the moist and mortal enemy of crisp.

Perhaps the most important difference between sandwiches and wraps is that in wrapdom, the mandate for bite consistency is less clear-cut. That's because bite variety is a natural outcome of wrap construction, which tends to create interior pockets of differential ingredient distribution.

Furthermore, while you can only approach a sandwich along one plane, you can eat a wrap from any angle, which means even if the fillings were layered perfectly, bite variety would still be easy to come by. (Want a different bite? Just rotate a few degrees.) The fact that bite variety comes so naturally to the wrap form tells us that both options merit places in the Eatscape.

FILLING AND WRAPPING TECHNIQUES: THE HIGH HORSE, THE SNAIL'S COCHLEA, AND EL MIXTEC

How best to fill and wrap a wrap? It mostly depends on whether you want bite consistency or bite variety. While I consider bite consistency to be preferable here as in sandwichdom, with wraps this is ultimately a matter of taste, not truth.

To impose consistency, you can try to layer and wrap ingredients with great care. But the easier way to do it is with a technique known as El Mixtec, where you chop and mix the fillings before they're wrapped, so they're already consistent throughout.

When it comes time to wrap the wrap, you have two basic options: The High Horse (figure 3.8) and the Snail's Cochlea (figure 3.9). The difference is mostly one of personal preference. The High Horse involves creating one high pile of fillings in the center of the wrap, then folding the wrap over the pile. It's the most common method of wrap engineering and produces pleasurable enough results, albeit with pockets of disproportion.

With the Snail's Cochlea, you spread fillings thinly across the entire surface, then roll the wrap tightly, like a window shade or scroll, to create a spiral effect inside. This increases bite consistency, because tighter circles prevent fillings from shifting so much during the wrapping. It also provides a different textural and visual experience, with ribbons of wrap wending throughout.

The High Horse

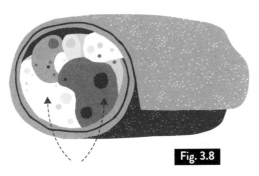

Fig. 3.8

These regions of disproportion can be remedied with El Mixtec.

The Snail's Cochlea

Fig. 3.9

THE VARIETY-CONSISTENCY SPECTRUM IN WRAPS

Here are the possible combinations of filling and wrapping techniques, ordered from the one that provides the most bite variety to the one that provides the most bite consistency. ("Standard" is the most common filling approach, the alternative to El Mixtec, where ingredients are layered without being mixed beforehand.)

1. Standard/High Horse

2. Standard/Snail's Cochlea

3. El Mixtec/High Horse or Snail's Cochlea (Once the fillings are fully mixed, the wrapping technique becomes a matter of personal preference only.)

WHEN TO POP CAPS ON YOUR WRAPS

When you fold in the end of a wrap in order to seal it, that is called "capping." Capping one end offers clear structural benefits, but capping the other end is purely aesthetic. (Most restaurants cap both, to keep fillings together and increase portability. But if you simply hold the wrap vertically, you can leave one end open without incident.)

In other words, double capping should only be done if you want to eat more cap. That's because any form of capping reduces the space available for fillings in the wrap interior. If the wrap itself is large to begin with, you may have space to spare, in which case double capping becomes a viable and pleasant option.

I happen to enjoy a few extra bites of pure wrap cap, both for its toothsinkability and its labyrinthine beauty. But I'll give it up in a heartbeat if I don't have the real estate.

WAFFLES AND OMELETS: STRUCTURES TO BEHOLD

Waffles and omelets are a perfect marriage of strength and beauty. They're engineered to be powerful and toothsinkable, to support the weight of toppings and fillings. But their brawn belies supple sides.

Waffles possess both stoic spines and delicate bread interiors, providing textural variety and extraordinary resistance to pressure for such light structures. Omelets offer a decadent ribbon of cheese encased in fluffy egg, with enough meats and/or veggies suspended throughout to make them hearty without threatening structural integrity.

In this section I'll pay homage to these classic structures, while also proposing methods for their improvement.

SCHOOLING ESCOFFIER

In his century-old tome *Le Guide Culinaire,* which established modern French cooking as we know it, Auguste Escoffier wrote:

The theory of the preparation of an omelette is both simple and at the same time very complicated, for the simple reason that people's tastes for this type of dish are very different—one likes it to be just done, and there are others who only like their omelette when it is extremely soft and underdone. The important thing is to know and understand the preference of the guest.

Can you believe this guy? He'd never survive on *Iron Chef,* I can tell you that much. There is no place for uncooked egg inside an omelet. It suggests sloppiness and reduces structural integrity. I've devised a much better way to incorporate runny eggs into the omelet experience, and I call it Escoffier's Bounty.

ESCOFFIER'S BOUNTY

If you want an omelet with the optimal runny-egg experience, cook your omelet through and eat it with a side of sunny side up eggs. Then dip bites of omelet into runny egg on a per-bite basis as you like.

You see, Monsieur Escoffier, it's one thing to "know and understand the preference of the guest." It's quite another to know and understand a superior alternative, then make it the preference of the guest.

OMELET ENGINEERING AND SANDWICHIZATION

I wish Escoffier's folly ended with his endorsement of underdone omelets, but I'm afraid it goes much further. The French style of omelet is lightly filled and rolled or folded in thirds, a suboptimal experience on several levels. When making, consuming, and judging omelets, keep in mind the following:

- All meats and vegetables should be mixed in with the omelet proper, not wrapped in the egg like it's some kind of bastardized burrito. This way, flavor from fillings comingles with the egg. Cheese, however, should be added after the egg has solidified, so the cheese forms a distinct layer. If it's allowed to melt into uncooked egg, it disturbs omelet consistency.

- Eggs are the connective tissue of the omelet structure. If you add too many fillings, the center will not hold. In theory you could add one tiny piece each of twenty different fillings, but then you can't tell what you're eating. That's why the ideal omelet has no more than three fillings in addition to cheese. (And two plus cheese is usually plenty.) This ensures a large enough quantity of each to be tasted but a small enough total to preserve eggtegrity.

- Omelets taste better when flipped in the air, without a spatula. Try it first over the sink. Make sure you have a well-oiled nonstick pan. It's all in the wrist. One success is worth many failures.

- The ideal omelet is folded in half to form a semicircle, not rolled or folded in thirds to form a cylinder, as in the French style. (See figures 3.10 and 3.11, and the recipe for Inside-Out Omelet Sandwich.)

The Problem with Omelets Folded in Thirds

There are two reasons why omelets folded in thirds (to form a cylinder) are inferior to those folded in half (to form a semicircle). First, to be able to fold an omelet twice instead of once (as in the cylinder), you need one with a larger diameter, so you have more real estate to work with. If you're using the same number of eggs as in a semicircle omelet, that means those eggs must be spread more thinly, which means less cohesion when fillings are present and less egg toothsinkability in some regions.

Furthermore, as the drawing below shows, you end up with a center that has three egg layers and edges that have only one egg layer, so the edifice is lopsided, with a large mound in the center. This makes the omelet less sandwichizable, because it refuses to lie flat on the bread.

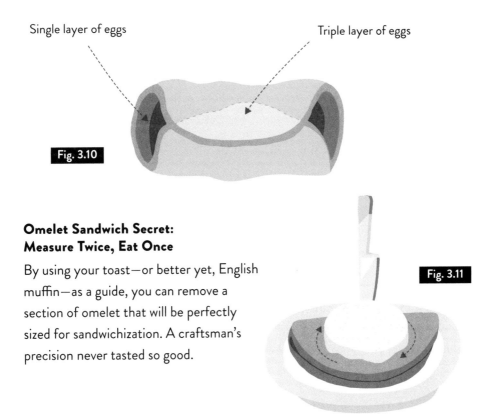

Single layer of eggs

Triple layer of eggs

Fig. 3.10

Fig. 3.11

Omelet Sandwich Secret: Measure Twice, Eat Once

By using your toast—or better yet, English muffin—as a guide, you can remove a section of omelet that will be perfectly sized for sandwichization. A craftsman's precision never tasted so good.

INSIDE-OUT OMELET SANDWICH

After you've traced and removed a portion of your omelet for sandwichization (figure 3.11), turn it inside out before placing it between the bread. This moves the cheese closer to your tongue, accentuating cheesy goodness, and fuses the bread to the omelet, improving structural integrity.

WAFFLE STRENGTH AND REINFORCEMENT

Because of the simplicity and ingenuity of the waffle maker, waffles are easier to make than omelets. But that doesn't mean they can't also be made more better.

The waffle's strength and beauty come from its hallmark perpendicular beams, which provide high SATVOR, textural variety, and resistance to pressure from toppings. As syrup and other forms of moisture seep in, the perfect waffle transitions slowly from crispy to chewy to tender, maintaining crisp in the areas left untouched.

But there are ways to turn a waffle into something even more powerful, by further reinforcing it with edible rebar not unlike the reinforcing steel rebar placed inside concrete. This not only strengthens the waffle but also adds flavors and textures that you used to need silverware to incorporate.

WAFFLES WITH BACON REBAR

This concept came from Eater Jeff Bentch of Houston, Texas. Pour half a portion of waffle batter into your waffle maker. Layer warm, crispy, fully-cooked bacon on top, and pour the rest of the batter over it. Close the waffle maker and use it as you would normally. The waffle will cook around the bacon, producing a waffle reinforced with bacon rebar.

I've taken Jeff's work a step further and done this successfully with frozen peanut butter cups, chicken fingers, and White Castle cheeseburgers, although in the case of that last one, your results may vary depending on your definition of success.

I should mention that I broke the handle off a weaker waffle maker while trying to close it around the chicken fingers. If you have the plastic-handled variety, slice your chicken fingers in half the long way first, to make them thinner. Whatever you put in there, it should be fully cooked in advance.

WAFFLES WITH PEANUT BUTTER SEALANT

Here's another way to imbue waffles with new physical properties. By spreading peanut butter across the top of multiple wells, you seal in butter and syrup, which meld with each other and the peanut butter to intoxicating effect. Plus it reduces syrup drippage, making it easier to eat waffles with your hands. (Eater Alison Ruggeri of Hailey, Idaho, and her sons, Junior Eaters Emmett and Owen, helped test this one. She reported back, "This idea is worth the price of your book.")

Alternatively, you can apply the sealant first, then drizzle syrup on top. This slows the absorption of syrup into the waffle and helps to ward off soggage.

IS IT BETTER TO SLICE WAFFLES ALONG SPINES OR ACROSS WELLS?

We've covered how to strengthen waffles. Now let's discuss how to destroy them, with someone who brings special expertise to the table.

Eater Isaac Gaetz is a licensed structural engineer in Chicago. When he's not sharing information with the rest of the Eatscape, he's building bridges and power plants. But just because he has some fancy degree doesn't mean he's always right. In fact, on this issue, I think he's dead wrong. Here are our respective arguments:

Cut Along Spines (Isaac)

1. Leaving the valleys unsullied allows us to transport a larger quantity of syrup and butter than is possible if the valleys are bisected. Slicing through the valley results in massive syrup loss.

2. Syrup is a viscous liquid, so it will naturally collect and pool at the lowest point of the waffles. If we slice through the hard waffle exterior at the base of the valleys, we expose the delicate inner waffle layers (strata) to direct submersion in syrup. The syrup will infiltrate the waffle and as a result, the layers of the waffle will begin to delaminate. Rapid deterioration of the waffle's structural load carry capacity is anticipated.

3. The waffle structure requires a continuous rib all the way around its circumference. As a quirk of this form, cutting through the waffle valleys will always require more cuts than cutting along spines (figure 3.12). This is an inefficient use of energy and wastes valuable waffle-eating time.

Cut Across Wells (Me)

1. The primary purpose of waffle wells is not to hold syrup, it's to provide textural contrast. The best way to maintain crisp is to keep your syrup reservoir separate and dip the waffle on a per-bite basis. This ensures perfect syrup levels in every bite and renders Isaac's first two points moot.

2. The spine is the source of the waffle's crisp and crunch. To stab a knife into it is to betray the waffle and to rob it of its most precious resource. When you slice the spine in half you turn a sturdy ridge into a delicate precipice, and force me to mix my metaphors in the process.

Fig. 3.12

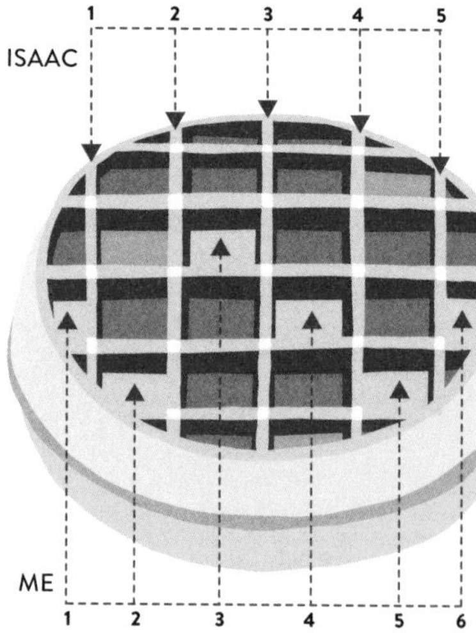

Cutting across waffle valleys requires more work than cutting along spines.

3. As to Isaac's point regarding the number of cuts, you know what's a waste of energy and time? Building beautiful, strong waffle spines, then tearing them down before you've had a chance to enjoy them.

WAFFLES ARE SUPERIOR TO PANCAKES, BUT THE PORKLIFT MAY CHANGE THAT

Isaac isn't always wrong. In fact, he gets a lot of things right. (See also his role in my tortilla chip breakthrough on page 50 in "Language Arts.") And while we don't agree on how to eat waffles, we do agree that they're better than pancakes.

"As a syrup delivery system, [the pancake] relies on a primitive sponging mechanism, which leads to rapid deterioration of its structure," says Isaac, totally taking the words right out of my mouth.

Conversely, "[the waffle's] cells allow for a much higher volume of syrup delivery, and don't require the waffle to substantially compromise its own structural stability in the syrup delivery process." In other words, pancakes have little crisp to begin with and turn to mush when confronted with syrup. Waffles have lots of crisp and can maintain it in the face of syrup, butter, fruit, and more.

However, I have devised a system to alleviate the soggage that so commonly befalls the pancake stack. I call it the Porklift.

The Porklift

Build a bacon lattice and place it beneath the pancake stack. The bottom pancake is elevated safely above the syrup pooling on the plate, leaving the bacon to bathe below it.

HOMEWORK

Let us culminate our engineering discussion with a dish that may best be described as "audacity on a plate." It is the Geodesic Breakfast Dome.

The GBD brings together many of the principles and foods in this chapter in a way that is delicious, revolutionary, and structurally impeccable. It is clearly the intertransformative breakfast edifice of the future. Soon it will be the only food available at the Buckminster Fuller Institute, Epcot Center, and everywhere fine geodesic domes are displayed.

Using the sketch below as a guide, build and eat a Geodesic Breakfast Dome. Add your own personal touches to the concept, and feel free to work (and eat) in groups.

The Geodesic Breakfast Dome

Layers of hash browns, cheese, and ham, with a scrambled egg core

Submit your homework to me at dan@sporkful.com.

4.
PHILOSOPHY
Sustenance and Existence

- Breaking down societal constructs that stand in the way of Perfect Deliciousness

- Time as a seasoning

- Reincarnating stale bread and old sandwiches

- Toast-crumb utilitarianism and the Majority of One

- The social contract of potluck dinners

- Socrates and snack mix ethics

- The Kama Sutra of Consumption

- Sun Tzu and the Art of Food Festival War

We begin with the words of Plato:

There will be no end to the troubles of states, or of humanity itself, till philosophers become kings in this world, or till those we now call kings and rulers really and truly become philosophers, and political power and philosophy thus come into the same hands.

Absorb and implement the lessons you learn here. Spread them throughout the Eatscape and bring more Eaters into our midst. Over time, we'll amass enough political power to implement our philosophy.

Of course, some will cling tight to cherished beliefs and refuse to join the flock. Our path can be rugged, steep, and lonely. But we forge on, taking inspiration from transcendentalist philosophers like Ralph Waldo Emerson. In his classic essay "Self-Reliance," he said, "Trust thyself," and famously wrote of those who try too hard to conform, "A foolish consistency is the hobgoblin of little minds."

Our minds, however, are as open as our mouths and as insatiable as our stomachs. We feed on discovery and innovation, pushing ever closer to the exalted heights of Perfect Deliciousness. In time we will claim our mantles as philosopher kings and put an end to suffering!

THUS ATE ZARATHUSTRA: EATER AS ÜBERMENSCH

What is objective truth? Philosophers can't even agree that we actually exist. What we call reality could all be a dream in the mind of a giant space turtle. But while this realization disturbs most people, Eaters welcome it.

Society is your blank canvas. Embrace the opportunity to construct new rules and belief systems to add meaning to your meals. Break the binds of convention and you'll stop *being* and start *becoming*.

DINNER, LUNCH, AND BREAKFAST—IN THAT ORDER

Why are some foods "breakfast foods" and others "dinner foods"? Is it because these foods are so inherently different from each other? Not so much. Look closely at the foods associated with certain times of day and you'll see that the difference is often one of perception, not reality.

At what meal would you typically eat a ham and cheese sandwich with French fries or potato chips? Probably lunch, maybe dinner. Why not breakfast? Ham, cheese, bread, and potatoes are common breakfast foods when reconstituted in the form of a ham and cheese omelet, home fries, and toast.

A steak screams dinner—until it's placed alongside eggs in a classic brunch combination. A slice of cake is usually reserved for the coffee hours of midafternoon or later evening—unless it's called a muffin, Danish, or donut.

How about fish? Sushi rolls make for a lovely breakfast, a lighter version of that classic morning combination of fish and starch, the bagel and lox. Sashimi is even better, because as leftovers, the fish holds up better than the rice.

As for dinner, eggs can be as glorious a protein delivery system as any meat. Savory oatmeals with ingredients like pesto and Parmesan make for excellent entrées, especially after an afternoon outside in the cold. And there's not a single moment in the twenty-four hours of the day when a peanut butter and jelly sandwich is not appropriate.

Eat the foods you want to eat when you want to eat them. Pay no mind to the small-minded who find it strange to start the day with a helping of steak and cake.

CHEDDAR GOLDFISH CEREAL

Here's a recipe that upends conventions about what foods should be eaten when. Cereal is essentially bread. Bread goes well with cheese. So why aren't there any breakfast cereals made with cheese?

Fill a bowl with cheddar Goldfish or a similar cheese-based cracker. Pour milk on top.

When you eat this dish for the first time, understand that the confusion your palate experiences is a natural reaction to its unshackling. Do not compare it to other cereals, which your brain expects to be sweet. Appreciate it as its own food, which defies categorization—and is oddly delicious.

TIME AS A SEASONING

Time is such an imperfect construct that every four years, the whole world agrees to pretend there's an extra day in the calendar that wasn't there before, just so our universe doesn't spiral out of control. There's no reason why time can't be manipulated to suit the Eater's desires, especially if you think of it not as an immovable force but as a seasoning, like salt or pepper, that you can add as you like.

Certain foods are greatly improved with time, especially cold ones, because cold dulls flavor. (That's why I oppose frosted mugs for beer.) Add a little time to foods just out of the fridge or freezer and they'll taste better.

Other foods don't necessarily improve with time but can take on new qualities that have merit in their own right. Day-old donuts and popcorn, for instance, lose crisp but gain a pleasant chewiness.

ICE CREAM WITH TIME

Ice cream tastes sweeter, richer, and creamier when it isn't frozen solid. It's also gentler on the teeth. Add some time and experience the difference.

YOU WILL NEED:

Ice cream

Time

INSTRUCTIONS

Scoop ice cream into bowl. Wait. When at least 10 percent of the ice cream has turned to liquid, consume.

TIME-ENCRUSTED ORANGE WEDGES

Peel orange and separate into wedges. Place wedges on a plate so that none of them are touching another, maximizing their exposure to air. Let sit for several hours, as a natural crust forms around the exterior of each wedge.

As you bite into one of these aged morsels, enjoy the contrast between its dry, crusty exterior and the cavalcade of juice within. Vary amount of time added to suit your taste.

ON DEATH AND REINCARNATION

What if death, too, is but a construct? Can spoiled foods be reincarnated? Can those going stale be brought back from the brink?

Philosophers have pondered this question for millennia, but only now, with the emergence of new technologies and techniques, has food reincarnation become possible. This notion is known as the Doctrine of Eaternal Recurrence.

TOAST PHILOSOPHY

The best way to reverse bread's demise is to toast it. But do not toast fresh bread—it's akin to embalming the living.

One who prepares toast is a "toaster." Too often, toasters make the mistake of considering only the crispy exterior of their creation. But toast should have softer interior strata as well, however fine.

If bread is toasted to the point that there is no interior textural variety whatsoever, it has been overtoasted. If that's how you like your toast, then you don't actually like toast. You like crackers. Eat those.

Some people may still choose to eat toast well done, just as they may choose to eat steak well done, even though that means they don't actually like steak either. People who like well done steak served on well done toast don't like anything at all. They are called nihilists.

THICK VS. THIN TOAST: PROS AND CONS

Toast comes in two categories—thick and thin—where the dividing line is three-quarters of an inch in thickness. Both have their place, but understand the pros and cons of each before deciding which to eat.

TOAST TYPE	PROS	CONS
Thick	More strata means it's capable of more textural variety in a single slice; greater margin for error, especially if the toaster likes toast medium well	Can throw off ratios, putting too much emphasis on toast when used for sandwichization or as a spread conduit
Thin	Can be magical when done well, because wisps of soft interior are so fine and delicate	Little margin for error, easy to overtoast

TOAST CRUMB UTILITARIANISM AND THE MAJORITY OF ONE

When buttering toast, you may move the knife between toast and butter more than once. Some toasters take extreme care to avoid returning butter to the fridge with remnants of toast on it. Their caution is misplaced.

Utilitarian philosophers believe that morality is largely defined as whatever will maximize pleasure and minimize pain for the greatest number of people. Everyone from Epicurus to John Stuart Mill agrees that the desire to seek happiness and avoid misery is natural.

That's why those toast crumbs adhere so readily to the butter. If you knew that remaining on the toast proper would lead to your demise at the hands of a hungry Eater, wouldn't you seek solace in butter's creamy bosom?

Once ensconced there, these toasty bits begin a metamorphosis into a delicate butter/toast hybrid that's wonderful when spread on a multitude of foods, including—you guessed it—your next batch of toast.

Permitting the crumbs to end up in the butter results in the most happiness for the greatest number of crumbs, and returning those butter-soaked crumbs to future toast results in the most happiness for the greatest number of Eaters. That the Eater eventually consumes the butter-soaked crumbs is evidence of what Thoreau called "the majority of one."

SEEKING PLEASURE (AND PEACE) BY BUTTERING BOTH SIDES OF YOUR TOAST

Society dictates that the butter be placed on one side of the toast and that the toast be eaten butter-side up. But when toast is eaten upside down, the butter is placed directly on the tongue to accentuate butter flavor. Of course, the true Epicurean may butter both sides, especially if the bread available is unworthy of a starring role. (Use an edge-hold method to keep fingers clean.)

One additional benefit of buttering both sides is that it may finally bring peace between the Yooks and the Zooks, factions long at odds over which side of the toast to butter. For more on this topic, read *The Butter Battle Book*, by the military historian Dr. Seuss.

REINCARNATED BAGEL

This amazing technique proves that even mortality is a state of mind. Begin with a defrosted, uncut, stale bagel. Preheat oven (or toaster oven) to 325°. Run hot water in your sink until it becomes very hot. Place bagel under hot water for 30 seconds, moving it around so it's coated evenly. Place it in oven for 5 minutes, directly on the rack. Remove and enjoy. Just be careful, when you slice it open: it'll be steaming hot!

RESUSCITATED PIZZA

Leftover pizza is never quite as good as the fresh variety, but it can still be enjoyable if you put a little extra effort into reviving it.

The problem with reheating pizza in an oven is that if you put foil or a baking sheet beneath it, the bottom crust never regains its crisp. If you put it straight on the rack, cheese oozes all over your oven and the crust gets burnt. Here's an approach that achieves better results on all fronts:

Heat a nonstick pan on medium heat. Place pizza in the pan, cheese-side down. Press down gently with a spatula to make sure all the cheese touches the pan. When oil starts to accumulate on the edges and all the cheese is hot, flip the slice and cook until bottom crust is crispy.

REVIVING A FORMERLY HOT SANDWICH WITH THE MICROWAVE/OVEN COMBO

Whether you're dealing with a grilled ham and cheese, a meatball sub, or a roast pepper and mozzarella panini, the issue here is the same: How do you heat both the sandwich interior and exterior so that the fillings are warm and melty and the bread is crispy?

If the whole sandwich is relatively thin and can be pried open, you can usually separate it into halves, with fillings divided roughly equally atop each piece of bread, and place both halves in the oven (open-faced) for a few minutes. Because the fillings are thin, you'll be able to heat them through without burning the bread. Just remove greens and delicate veggies beforehand, or they'll turn to mush in the heat.

But if the fillings are especially dense and/or the sandwich is fused shut, that won't work. So what to do?

If you microwave the whole thing for long, the fillings are heated but the bread is destroyed. If you put it in the oven, the bread will likely burn or turn rock hard before the interior is warmed through. Bringing it to room temperature helps, but it often doesn't fully solve the problem, and you may be hungry now.

Here are your options:

RESURRECTED FORMERLY HOT SANDWICH (WHEN IT CAN BE DISASSEMBLED)

Let's start simple, with a chicken parm sub that's been in the fridge, where the bread can be safely pried from the meat . . .

Separate fillings from bread. It's fine if some cheese stays with the bread. Heat bread in oven and fillings in microwave. When everything is warm, reassemble the sandwich. Consider alternate layering if you see opportunities for improvement.

RESURRECTED FORMERLY HOT SANDWICH (WHEN IT'S FUSED AS ONE)

Now things get more complicated. Let's consider a meatball sub that's been in the fridge for a day, so the cheese has fused the meat and bread into one structure . . .

Slice down middle of sub about three-quarters of the way through meatballs and pry them open. Do not sever hinge.

Microwave sandwich just enough to get meat a little warm. Bake in oven until bread is crispy and meat is hot.

Slice meatballs open so they heat faster

RESURRECTED FORMERLY HOT SANDWICH (WHEN ONLY A MICROWAVE AND POP-UP TOASTER ARE AVAILABLE)

This technique is especially helpful in office kitchens, where equipment may be limited. Open sandwich as much as possible, separating it into two open-faced halves if you can, and microwave it until fillings are warm through. Turn on pop-up toaster and lay sandwich across the top as shown.

ETHICS AND THE REPUBLIC

We've covered philosophical concepts as they relate to individual foods and scenarios. But when people come together to form societies, eating becomes more complicated. One Eater's meals affect another's, and we must seek ways to make the world a more delicious place for everyone.

THE SOCIAL CONTRACT OF POTLUCK DINNERS (OR, THE SOFT SOPHISTRY OF LOW EXPECTATIONS)

In any social contract, people give up certain freedoms in order to ensure others. The problem with most social contracts is that they lack ambition.

Is it really so impressive to get people to give up the freedom to murder just so they themselves don't get murdered? (Okay, maybe we're still working out the kinks on that one, but still, it feels like a low bar.)

We can do better. A higher-level social contract is in order, one that binds us together and unites our fates, while still letting us all pursue deliciousness in our own ways. It is the Social Contract of Potluck Dinners.

At a potluck, Eaters give up the freedom to dictate their whole meal in exchange for the freedom to take other Eaters' food, relief from the pressure of having to act as sole cook for a large group, and a unique shared experience that can enrich the entire Eatscape.

In this miniature society, there are rules. Sometimes rules are inconvenient. I'm sure there are times you wish you could just drive through a red light or annex the nicer house across the street. But you don't, because although the rules are imperfect, they protect you, and they form the basis for the society in which we live.

In the case of a potluck, the most important rule is that participants must not know what anyone else is bringing and must not have any restrictions placed upon what they might bring. The philosophical underpinning of this decree is simple enough—most of the word "potluck" is "*luck*." To tell people what to bring is to remove the luck. If you want to have friends over and give them specific instructions, that's fine, but then it's not a potluck. It's just a dinner.

True, you may end up with ten apple pies or a whole lot of potato salad. If so, you were unlucky. But a potluck society isn't some socially engineered utopia—it's more of an artist's colony. Some days, everyone paints the same apple. Sometimes dishes will clash or seem redundant. But when serendipity prevails and potluck foods fall perfectly into place, it creates a glorious tapestry, made more powerful because each Eater has placed trust, and stomach space, in the hands of the others.

FREE RIDERS AND THE POTLUCK

In any society, there will always be those who don't pull their weight. These people take more than they give, upending the social contract and sowing resentment among the populace.

At what point has the Social Contract of Potlucks been broken, and how should such breaches be handled?

At the very least, potluck participants must prepare something themselves, and it must be a good-faith effort to achieve the highest level of deliciousness that particular Eater is capable of reaching. That may just mean opening a box and a can and mixing the contents together, but the effort and desire must be there.

Bringing purchased, pre-prepared food to a potluck is a violation of the social contract, unless it comes with a truly exceptional extra touch or, failing that, a great story. (Positive contributions to the social part of society can also weigh in an Eater's favor.)

If someone's efforts are truly lacking, the recourse is obvious—the person should be placed on probation and offered counseling. Repeat offenders are not invited back.

THE JUSTICE OF PERFECT DELICIOUSNESS: A SOCRATIC DIALOGUE

Philosophers engage in dialogues with each other to test hypotheses and determine what rules should govern the Eatscape. Debates over issues like the meaning of justice, the role of government, and the nature of truth have continued through the ages. Many of Plato's famous works are actually dialogues in which he himself is absent and Socrates does most of the talking.

Let's take time now to study a dialogue between philosophers on the ethics of snack mix consumption.

Persons of the dialogue:

SOCRATES, who always wins

THRASYMACHUS, who is the Washington Generals
of Socratic dialogues

GLAUCON, who loves pretzels and is the host

ADDITIONAL GUESTS

*The philosophers arrive at the home of GLAUCON, who lives at the Port
of Piraeus. He greets them warmly and puts out a large bowl of snack mix
consisting of pretzels, peanuts, bite-sized bagel chips, and a square breakfast
cereal made of crosshatched fibers, from which they all readily partake.*

*SOCRATES picks out a handful of peanuts and bagel chips from
the mix, which draws THRASYMACHUS's ire.*

THRASYMACHUS: I see, Socrates, that you believe it permissible to
cherry-pick your favorite ingredients from a snack mix.

SOCRATES: I do. Should I believe otherwise?

THRASYMACHUS: You should. The snack mix is a delicate balance of
components. To target certain ones is to deprive others of the desired
experience. It is unjust.

SOCRATES: So do you believe, then, that the purpose of a snack mix is to
bring together ingredients in a proscribed ratio?

THRASYMACHUS: I do.

SOCRATES: And would it not stand to reason, then, that the Eater should
compose each bite in that ratio?

THRASYMACHUS: It would.

SOCRATES: How then should the Eater ascertain the proscribed ratio?

THRASYMACHUS: Well, I think it plain to see, Socrates. Survey the mix and judge the proper ratio based on its appearance.

SOCRATES: But, Thrasymachus, do not different components have different sizes and weights, such that some components may be found in greater abundance at the top or bottom of the mix?

THRASYMACHUS: Most certainly.

SOCRATES: And is it not also true that the mix maker's technique may result in additional disturbance and ratio inconsistency from one stratum to the next?

THRASYMACHUS: It cannot be denied.

SOCRATES: So we agree that the proportions of the top layer are ever changing. Thus, is it not also true that Eaters will partake in different ratios at different points throughout the mix-eating process, and thus continually alter proportions further?

THRASYMACHUS: To suggest otherwise would be foolish.

SOCRATES: We are agreed, then, that it's impossible to ascertain the precise ratio of mix components throughout the bowl, and that the ratio on top may not reflect the ratios beneath. Some might suggest that the mix maker counter this concern by binding the ingredients into clusters of consistent proportions, but even Peloponnesians know that this binding would create a cohesive new food and not a mix. So do you agree that there is no way to impose bite consistency throughout an entire bowl of snack mix, even if it were desired?

THRASYMACHUS: I have no reason to speak contrariwise.

SOCRATES: Good. Furthermore, is it possible that, if I select only peanuts and bagel chips from a snack mix, I am shifting ratios closer to another Eater's ideal?

THRASYMACHUS: Verily.

GLAUCON [*interjecting*]: Indeed, I like pretzels and am less fond of peanuts. I do not object to the Socratic method, as it increases pretzel accessibility.

SOCRATES: Thank you, Glaucon. Have a pretzel.

[*SOCRATES passes the bowl, and GLAUCON is much pleased.*]

THRASYMACHUS: Socrates, you are a sillybilly. What if I quite enjoy peanuts and bagel chips? I would consider it unjust to arrive at a snack mix to find all of those ingredients eaten.

SOCRATES: Well, Thrasymachus, a moment ago you agreed that attaining true bite consistency throughout a snack mix is impossible. Do you agree that if something is not possible, it should also not be preferable?

THRASYMACHUS: If not a shred of possibility exists, it is folly to prefer it.

SOCRATES: Well said. So the natural preference in snack mix consumption is for bite variety. Indeed, the ability to compose new and different bites throughout the eating experience is a central benefit of the form. The term "mix" refers to not only the method of preparation, but also the method of consumption. The very reason that components of a mix are never bound together is to offer the Eater a range of possibilities.

[*KANT enters.*]

KANT: Are you guys talking about snack mix?

GLAUCON: But of course. Join us.

KANT: Equal access to and distribution of snack mix ingredients is a categorical imperative.

[*KANT picks out one of each mix component and arranges them in his palm Germanically.*]

KANT: Subjective taste is irrelevant to this question, as it is to all issues of morality in eating. Perhaps, as you have said, Socrates, the natural ebb and flow of mix ratios is universally accepted and understood. But pure practical reason dictates that willful disfiguration of mix proportions is immoral.

[*Just then, a voice calls from the bathroom. It's NIETZSCHE. He's been lying on the bathroom floor. Everyone thought he was asleep.*]

NIETZSCHE: Oh, Kant, you're always worried about the weak.

[*NIETZSCHE stumbles into the living room.*]

NIETZSCHE: The Eater is a great philosopher king! His drive to consume the perfect combination of snack mix ingredients is the true manifestation of the will to power. The Eater has no responsibility for the experience of the lesser being, who would sit on the Eater's head like a dwarf and make it harder to chew! Besides, here we sit beneath a gateway. The name of the gateway is "This Snack Mix." From the gateway, all snack mixes run backward for eternity. Must not whatever combinations can exist have already existed? And therefore, too, must they not be destined to eternally return? Thus, why should any ingredient be too precious to devour by itself or with others as the Eater alone deems fit?

SOCRATES: I can't believe I'm saying this, Fred, but I don't completely disagree with you.

[*Suddenly HOBBES, LOCKE, and ROUSSEAU rush in.*]

HOBBES, LOCKE, and ROUSSEAU [*in unison*]: Not so fast!

NIETZSCHE [*groaning*]: Oh great, here come Nasty, Brutish, and Short.

HOBBES, LOCKE, and ROUSSEAU [*in unison*]: We disagree on quite a number of details, but we agree on the basic notion that without some form of social contract, society cannot function. He who takes selfishly from the snack mix one day will be victimized by another's selfishness the next. Eventually, society will spiral out of control and nobody will even make snack mix anymore, because they'll know that as soon as they make it, it will be destroyed. When one eats snack mix, one implicitly enters into a social contract under which he gives up the freedom to cherry-pick ingredients in exchange for the freedom to partake of the bounty at all.

GLAUCON: I wonder whether randomness has a role in this conversation.

ARISTOTLE [*bounding up the stairs into the scene and plunging his fingers into the bowl of snack mix*]: Surely it does. Some events are unknowable. Snack mix consumption is a game of chance. You put your hand in and hope the odds are ever in your favor . . . Oooh, bagel chips!

SOCRATES: What is the meaning of justice?

[*The room falls silent. THRASYMACHUS drops a peanut.*]

SOCRATES: Kant, you say that subjective taste is irrelevant. But in doing so you force the Eater to accept the mix maker's ratios, regardless of their merit, the Eater's tastes, or the preferences of others. That's not free, just, or reasonable. Hobbes, Locke, and Rousseau, I've grown tired of your slippery slopes. We're here in ancient Greece, walking around in bedsheets fashioned as garments, and yet even we can manage to create and share snack mixes from which many Eaters cherry-pick ingredients, without the fabric of civilization coming undone. Aristotle, if you consider snack mix consumption a random act, that is your decision. But the true Eater does not leave to chance that which can be made more delicious by choice.

[*The room remains quiet. SOCRATES is on a roll.*]

SOCRATES: Indeed, there is no greater justice than deliciousness, and the pursuit of that justice requires that one enjoy a snack mix to the fullest possible extent. Doing so means composing bites as the Eater desires them. The question here is, when cherry-picking is taking place, can all Eaters share equally in the benefits of just deliciousness?

I have no doubt that they can, and this leads me to my Theory of Snack Mix Forms.

Nietzsche's analogy to the eternal path was not far off. There is only one true snack mix, the Great Snack Mix in the Sky, which flows endlessly through the vast trough of time. From that mix, every conceivable bite can be composed at once, and no ingredient is ever lacking.

But when we eat snack mix, as we have done here today, we partake not of that most pure mix, but of a particular representation of it. These representations may vary, but when we eat them, we are all seeking to know and taste that highest Form, that most delectable reality, the one true mix. Each snack mix experience is another step in the same endless journey, not a discrete moment in time independent of the others. As with any long voyage, some steps may bring you closer to your destination while others may bear less fruit, or pretzels. Some days you may arrive at a snack mix that has been cherry-picked to oblivion, but over time and with persistence, you'll move ever closer to that Great Snack Mix in the Sky. Indeed, it is this most just pursuit of deliciousness that is the defining characteristic of the Eater.

[*SOCRATES is finished speaking. The room is quiet. The only sound is GLAUCON's faint pretzel-crunching. THRASYMACHUS blushes. HOBBES, LOCKE, and ROUSSEAU quietly put their hats on and show themselves out.*]

KANT: You guys know how to get to Prussia?

ARISTOTLE: I'll show you.

[*ARISTOTLE and KANT leave together. Nietzsche returns to the bathroom floor.*]

GLAUCON: Let's go for a walk, Thrasymachus. We're out of pretzels.

[*Exeunt.*]

EASTERN PHILOSOPHIES

Eastern philosophies teach us to transcend the everyday, in order to open new worlds of consciousness that exist within us. As Guru Ravi Shankar has written, "When the mind is free from all impressions and concepts, you are liberated. The state of nothingness is called Nirvana . . . When you go deep inside you, layer after layer, that is nirvana. It is like peeling an onion! What do you find in the center of an onion? Nothing!"

This realization isn't depressing, it's liberating, because all those onion layers were just clutter and distraction along your journey. Now that you've fried them up and eaten them, you can experience Nirvana.

KAMA SUTRA 101:
Types of Melon Union According to Dimensions and Force of Desire

Centuries ago in ancient India, Vatsyayana wrote *The Kama Sutra,* a Sanskrit pleasure manual that spawned a long lineage of scholarship to which this work proudly adds.

We Eaters often refer to the Kama Sutra of Consumption, which governs many gastronomic pleasures across the Eatscape, including melon eating. Depending on the type of congress desired and the force of desire, the Eater may seek different styles of melon union as follows:

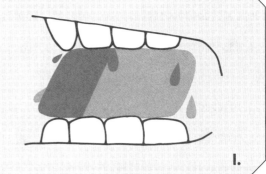

When an Eater eats melon squares, rectangles, or trapezoids cut to provide flat sides that lie pleasantly flush against the teeth, it is called "horizontal congress."

I.

When an Eater eats melons cut into balls, which possess a certain aesthetic beauty but are difficult to keep on a plate, it is called "stalking the siren."

II.

When an Eater eats melons cut into large wedges with the rind attached, making them easy to hold in the hand but messy on the face to eat, it is called the "marriage of convenience."

III.

When an Eater eats melons cut into very narrow wedges, so they're less substantial but can fit in the mouth and be eaten by hand without facial mess, it is called the "swallow's circumstance."

IV.

When an Eater eats melon cubed to include a bit of white rind to add a pleasing touch of complementary tart, it is called "congress with the courtesan."

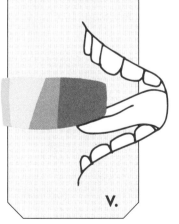

V.

KAMA SUTRA 201:
The Sandwich Embrace and Kiss

This section of the Kama Sutra of Consumption governs sandwich grips and bites.

When an Eater presses down on the top of a very tall sandwich to flatten the layers and make it mouth-ready, it is called the "ironing of the sheets."

I.

When an Eater arranges the hands like a net around the back of a sandwich to restrain disobedient fillings, it is called the "safe word."

II.

When an Eater nibbles the perimeter of a sandwich to trim fillings that protrude beyond the bread boundary, it is called "pruning the hedges."

III.

KAMA SUTRA 301
Mouth Congress

This section of the Kama Sutra of Consumption discusses ways in which various foods are eaten.

When at the halfway point of eating a coated ice cream pop, an Eater skips ahead to the bottom two corners, because they're the two best bites and they've reached the perfect level of meltiness at that precise moment, it is called "seizing the sparrow."

I.

When an Eater wraps food such as bacon strips or string cheese strands around the finger with the desire of kissing it, it is called the "twining of the creeper."

II.

When the Eater places an onion ring in the mouth so that its sides align with both sets of molars at once, it is called the "congress of equal proportions."

III.

When the Eater folds an onion ring in half and places it along the molars of one side of the mouth, it is called the "congress of crunch concentration."

IV.

THE YIN-YANG OF KETCHUP AND MUSTARD

At the core of the concept of yin-yang is the notion that throughout nature, forces that seem contradictory are often interconnected. Such is the case with ketchup and mustard, condiments so often served—but less often eaten—together.

In truth, ketchup's sweetness and mustard's sourness can make for a great pairing. Squirt them on your plate in a yin-yang formation and you can alter their ratios on a per-bite basis as desired.

WHAT WOULD SUN TZU EAT?

Sun Tzu is the military general and thinker whose ancient Chinese text, *The Art of War,* is still taught in business and military schools today. It's also applicable to our studies, because while we all wish the Eatscape was an entirely peaceful place, we know that's not the case. Whether we're battling the Buffet Master, as described in "Business and Economics," or gravity, as described in "Engineering," we will always have adversaries in the Eatscape. Facing our foes is the only way to clear the path to Perfect Deliciousness.

Using Counter-Siege Fortification to Build a Better Mashed-Potato-and-Gravy Vessel

Sun Tzu wrote, "The worst policy of all is to besiege walled cities." Of course, that means if you want security, a walled city is the way to go. That being said, he who builds a great citadel also knows best how to destroy it. This mashed-potato-and-gravy fort will keep gravy secure until *you* decide to tear it down.

Eat the outer perimeter first, dipping each bite into the gravy well and taking care to preserve the integrity of the empty moat, which will catch gravy runoff when you begin step two—tearing down the tower. Once you've dismantled the perimeter and central structure, the empty moat becomes a gravy boat. Combine remaining mashed potatoes and gravy as desired.

This is where you start eating. Remove bites and dip into gravy as desired.

Make sure you keep the inner moat wall intact during the first phase of destruction.

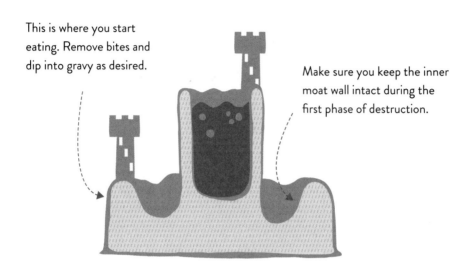

Eat the gravy tower second. As the walls come tumbling down, the moat will contain runoff.

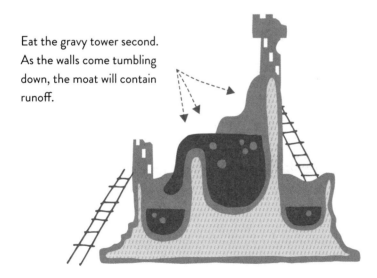

THE FOOD FESTIVAL FOG OF WAR

Invading a food festival can sound like a good idea when you see the intelligence on paper, but once inside, you may find yourself in a quagmire, suddenly aware that few food festivals are as good as they sound. That's because:

- There are usually more options than you will have time or stomach space to sample, but intelligence on where best to use your resources can be unreliable.

- The portions are often small and the lines at the best stands are long. This means that you get a small plate of food and eat it, then by the time you make it to the front of the next line, you're just as hungry as you were before you ate the last plate of food. Your stomach feels like it's on a hamster wheel, with satiety always just out of reach.

- Even when portions and line length are favorable for Eaters, purveyors still have to overcome the fact that it's nearly impossible to re-create a good kitchen in the middle of a farm, field, park, or parking lot. Quality suffers.

What does Sun Tzu recommend when faced with such a scenario? His timeless tips on how to win a war will help you win a food festival:

- "He will win who knows when to fight and when not to fight." Before deciding whether to go at all, research likely attendance, food-stand density, and venue terrain. If maneuvering will be too difficult, victory will be impossible, and you should not engage in battle.

- "The general who loses a battle makes but few calculations beforehand." Learn as much as you can about each stand before arriving—location, menu, product scarcity, and prices. Get there early to gather more intel and beat crowds.

- "He wins his battles by making no mistakes." Avoid bad decisions by scoping out the food a stand is serving before getting in line. If you see something on a stranger's plate that looks promising, ask for details.

- "If [your opponent] is taking his ease, give him no rest." Watch lines at top stands. If one suddenly shortens, move in for the kill. And remember, "He who is skilled in attack flashes forth from the topmost heights of heaven."

ZEN AND THE ART OF THE TAO OF EATING

While Eastern philosophies and religions have contributed a great deal to the Eatscape, I must express my concern that some of their teachings run counter to the Eater's way of life. I call your attention to two principles on which Buddhism and Taoism generally agree:

1. Resisting change is not only stressful but also pointless.

2. Constantly wanting things is bad.

How can we make the world a more delicious place without fighting against those changes that would push us toward meal mediocrity? How will we fuel our movement without the insatiable craving for more food? And if this is what Chinese Buddhists truly believe, why is their take-out food constituted to make sure we're hungry again two hours later?

On the other hand, I suppose it's good that there are sectors of society where the changes we Eaters advocate will meet less resistance. If a few hundred million people won't get in the way and won't try to take our food, I think we'll be able to coexist just fine.

HOMEWORK

This chapter was a lot to digest. I wouldn't blame you if you put this book down now and stared blankly at the wall for a few days. In time, however, I hope you'll return and forge ahead. This knowledge hasn't weakened you, it has made you stronger. Do not squander that gift.

Here's your essay assignment . . .

If Buddha, Sun Tzu, and Vatsyayana entered the preceding Socratic dialogue on snack mix ethics, what would be their respective positions? Explain how you arrived at your conclusions.

Submit your homework to me at dan@sporkful.com.

5

BUSINESS & ECONOMICS
Profit for the Palate

- Maximizing deliciousness per dollar

- Defeating the Buffet Master

- Sunk costs, diminishing returns, and rational eating

- Supply and demand at airports, concerts, and sporting events

- Drink specials and loss leaders

- Fad diets and Ponzi schemes

- The craft beer bubble

- Food marketing gimmicks

- Keynesian stomach economics and the ideal bowl of cereal

Your stomach space is a limited resource, and the money in your bank account probably is too. On the journey toward Perfect Deliciousness, we Eaters pursue maximum deliciousness at minimal cost, in order to generate the greatest possible profit for our palates.

But how best to bring home the most bacon? Adam Smith, the father of modern economics, explains in *The Wealth of Nations*:

> *It is not from the benevolence of the butcher, the brewer, or the baker that we expect our dinner, but from their regard to their own self-interest. We address ourselves not to their humanity but to their self-love, and never talk to them of our own necessities, but of their advantages.*

So let me talk to you of *your* advantages. In this chapter, we'll discuss a hallmark subject of gastroeconomics—the all-you-can-eat buffet—and identify strategies to maximize gains in the face of the dreaded Buffet Master. We'll also cover business principles like cost-benefit analysis and supply and demand as they pertain to everyday eating, show how to exploit market inefficiencies, and debate whether Eaters always act rationally. Finally, we'll use the work of Smith and John Maynard Keynes to determine when to intervene to attain the best possible cereal outcomes.

Put the following economic principles into practice, and you'll conserve both physical *and* fiscal resources. Bottom line: It'll be good for your bottom line.

SUPPLY AND DEMAND

The relationship between supply and demand has many inefficiencies, which businesses seek constantly to exploit in their favor. Turn the tables on these unscrupulous practitioners and you'll always have the food supply to satisfy your stomach's demands.

DEFEATING THE BUFFET MASTER

In an efficient market, prices go up and down to balance supply and demand. But at an all-you-can-eat buffet, the price is fixed and supply is limitless. You can demand as much food as you want without paying a penny more. A hefty profit would seem inevitable, were it not for a formidable business adversary: the Buffet Master.

This diabolical overlord is your competitor in a zero-sum game. The more you fill your face, the less he fills his coffers—but he won't go down without a fight. He knows full well the weakness in your business plan: You only have so much stomach space, plate space, and patience. The more cheap food he can get you to consume, and the slower he can get you to consume it, the better for his bottom line and the worse for yours.

So the Buffet Master stacks the deck against Eaters, using the dark arts of chafing dish arrangement and line choreography. (It's no coincidence that buffets thrive in casinos, another place where the odds always favor the house.) He builds the buffet full of obstacles, traps, and distractions designed to keep you from eating his most treasured culinary delights (figure 5.1). He puts the filler at the start of the buffet, and each time you sink your teeth into a roll, he cackles. When you line your plate with rice and salad, he lines his pockets. And as you sip a spoonful of soup, he strokes his cat, Sterno.

If you make it to the hallowed meat carving station with stomach space, plate space, and sanity to spare, you'll find the most skilled of the Buffet Master's minions. Each carver is a steely-eyed Cerberus trained to minimize your access to succulent prime rib and other roasted joy. They shave impossibly thin slices that take up a lot of plate real estate without actually constituting much meat, so you'll have to be patient. Demand slice after slice until your plate is filled to your meaty satisfaction, paying no mind to the crowd stacking up behind you.

If you sense disapproval from employees or patrons, simply remind them: It's nothing personal. It's just business.

TIP The red tint of many heat lamps makes meat appear rarer than it actually is. The rarest meat will always be in the middle of a roast, so don't be shy about asking for a center cut.

GAME THEORY AND BUFFET STRATEGY

To be clear, your strategy at a buffet should not be to eat the most expensive foods simply for the sake of it. Your primary goal, as always, is to maximize deliciousness at minimal cost, which means you may eat items of moderate value if they're especially tasty.

That being said, nobody likes the taste of getting ripped off. So know what you're paying for the buffet, and make sure you consume a meal that, eaten à la carte, would have cost more. If you do, you can claim a noble victory.

To maximize deliciousness at an all-you-can-eat buffet, follow these steps:

1. Avoid arriving when the buffet first opens, when they're serving last night's leftovers, or right before closing time, when the kitchen stops refilling some stations.

2. Sit in a location with a good view of fresh food coming from the kitchen and line lengths at the buffet.

3. *Survey the entire buffet before taking any food.* This is the most crucial component of good buffet strategy. Teach your children.

Typical Buffet Setup with Obstacles Highlighted

The Buffet Master puts cheap filler at the start of the buffet, hoping you'll gobble it up, and places high-value items like prime rib at inconvenient locations, so bottlenecks form and patrons become discouraged by long lines.

Fig. 5.1

4. On your first pass, take very small amounts of everything you want to try. Do not feel obligated to fill every corner of your plate. If something turns out to be unappetizing, do not eat a single bite more of it. No further stomach resources should be spent on subpar foods, but by taking very small amounts to begin with, you reduce waste.

5. Don't drink too much of anything, especially carbonated and/or caffeinated beverages, which are cheap appetite suppressants. The Buffet Master may be stingy with steak, but he's benevolent with beverages. He knows that if your stomach economy is bloated, you won't be able to take advantage of changing market conditions, like a fresh rack of lamb appearing at the carving station late in the meal.

6. On your second pass, fill your plate with prodigious piles of the most delectable dishes from the first pass. To get seconds without waiting in line again, try the Hummingbird Technique (figure 5.2) or Riding the Wave (figure 5.3). Just remember to bring your used plate back when you go, to send a signal to other diners that you already waited in line once.

7. Continually monitor the buffet from your table, watching for reinforcements of key items and unusually short lines at prime stations. When an opportunity arises, strike decisively. Keep in mind, however, that the Buffet Master's minions bury new offerings under old ones. Dig deep to get the fresh stuff.

8. With the third pass you strike a death blow to the Buffet Master, as you take more of the best items. On the way back to your table, you may wish to stop near a security camera, stare at the Buffet Master through its lens, remove a slice of prime rib from your plate, and gnaw it suggestively. He hates that.

9. The fourth pass is for dessert. Stomach resources are now scarce, so work with your dining partners to get one of each promising option and share them all. Desserts tend to be cheaper, so if you make them your buffet focus, you're getting less value.

The Hummingbird Technique (or, The Hover-and-Dart)

This is a quick-strike approach suitable for any spot in the line. Begin by hovering near the food you want. As the line moves and one person steps forward, the person behind them may take a split second to move up, which creates a small gap in the line. That's when you dart.

As you dart, be careful not to step fully into the line. Keep your feet outside the path of the line and lean into the gap, which shows that you're not cutting in, just making a quick tactical strike.

Fig. 5.2

Riding the Wave

This technique takes advantage of the natural ebbs and flows of the buffet line. There will always be some stations that slow the line down, in part because the Buffet Master wants it that way (as discussed earlier), and in part because certain foods just take longer to get on the plate.

There are often gaps in the line just beyond these bottlenecks, creating a pattern in the line similar to waves. By watching the tides you can find openings and ride them to the shores of satiety. This approach only allows you to access the buffet from certain points, but you can do so more comfortably than with the Hummingbird.

Fig. 5.3

SUNK COSTS AND BAD BUFFETS

Even the most seasoned Eater will occasionally sign up for an all-you-can-eat buffet that turns out to be disappointing. If you find yourself in this situation, understand that the money you're spending is a sunk cost—it's gone, no matter how much you eat.

If the food is great, of course, you want to eat as much as you can. But there's no value to be found in eating more bad food, no matter what you've paid for it.

Walking away isn't easy. Economists have shown that part of the sunk-cost effect is that the more people invest in something like a buffet, the more they believe it'll work out well and the more likely they are to stick with it, in spite of evidence to the contrary. In truth, if the food is bad, consuming more only magnifies the pain. So when you're in that situation, don't cut more chicken. Cut your losses.

BUFFET FOODS THAT CALL FOR CAUTION

- **FISH** It's generally more delicate, so cooking huge quantities of it well is difficult. Plus brunch buffets in particular are often a way for restaurants to get rid of fish before it expires.

- **FRIED FOODS** It depends on how they're served. If they're piled densely, their collective heat can lead to condensation, which reduces crisp. (See page 21 in "Physical Sciences" for more on this issue.) Look for pieces with sufficient air flow around their exteriors.

- **LAMB** It can be great, but it's not always so popular. Watch out for foods that may have been sitting out longer due to a lack of demand.

- **STEAMED VEGGIES** They're rarely good, and if they are, it's because they're soaked in butter. That's a technique you can replicate at home. And if your goal is to eat healthy, what are you doing at an all-you-can-eat buffet in the first place?

AIR-TRAVEL EATING: FINDING DELICIOUSNESS WHEN EVEN THE AIR TASTES BAD

Life isn't always a buffet. Let's look at the inverse situation, in which supply is low, demand is high, and prices are outrageous.

Airport terminals and airplanes are sealed and often crowded environments, with many people and limited food options. What's there is usually expensive and bad, in part because this closed market lacks competition and in part because cooking in an airport or airplane is hard. There's little space, so meals are often made elsewhere and brought in. Restaurants can't get security clearances for the best chefs, 97 percent of whom are nonviolent drug offenders. And they have to deal with that air.

Indeed, the air in air travel is the greatest enemy of a decent meal. The *New York Times* explains:

> *As the plane ascends, the change in air pressure numbs about a third of the taste buds. And as the plane reaches a cruising altitude of 35,000 feet, cabin humidity levels are kept low by design, to reduce the risk of fuselage corrosion. Soon, the nose no longer knows. Taste buds are M.I.A. Cotton mouth sets in. All of which helps explain why, for instance, a lot of tomato juice is consumed on airliners: it tastes far less acidic up in the air than it does down on the ground.*

I strongly suspect that this awful air also seeps into the airport proper and infects air-pressure systems in airport bars, which is why it seems impossible to get a decent draft beer while waiting for a flight.

Of course, maximizing palate profit doesn't just mean cashing in on the good days. It also means keeping your head above water during tough times. That's why savvy Eaters always travel with their own food. It doesn't even have to be that good, or that cheap, to be a vast improvement. And if you plan well, you can actually capitalize on cabin conditions, because while taste perception decreases, the white noise of the engine has been found to increase crunch perception. So bring the crunchiest food you can find and GO TO TOWN.

You won't just be eating more better, you'll also be doing the whole Eat-scape a favor. According to Adam Smith, if we all bring our own food and stop buying airport food entirely, it'll be in purveyors' self-interest to improve quality and decrease prices. So let's do that.

IMPROVING YOUR IN-FLIGHT MEAL

If, for some inexplicable reason, you've failed to pack yourself a meal and some snacks before embarking on your journey, there are things you can do to improve your prospects. Keep these tips in mind:

- As is generally the case with prepackaged, highly processed foods, saucy pasta dishes fare better than large pieces of meat.

- The kosher option is recommended, because the meat will be marginally higher in quality and specialty meals are often served first. Vegetarian meals are another good way to get food sooner while avoiding questionable meat.

- Never order fish. Fish lives below sea level. You're five miles aboveground. Fish are meant to move horizontally, which means they can be transported across land to your plate. But they should never be eaten that high above their home. (By contrast, I've wondered whether birds would be more delicious on airplanes, because you're already so close to their

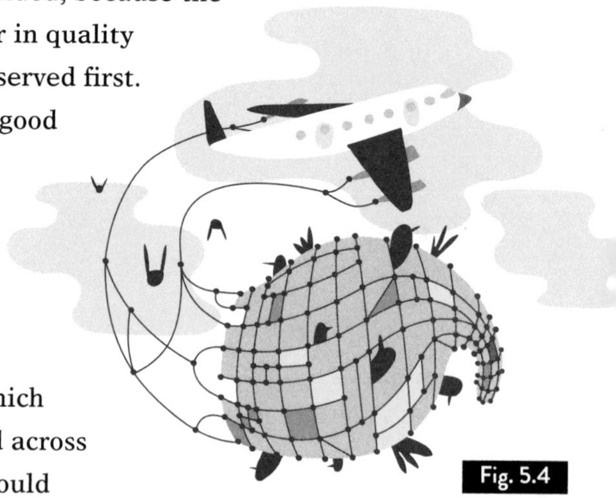

Fig. 5.4

Airplane Equipped with Bird Trawlers

Airplanes already kill a lot of birds. This new technology would ensure that their deaths are not in vain.

natural habitat. Perhaps aircraft could be equipped with nets, not unlike fishing trawlers, and airplane food could be harvested in flight. See figure 5.4. More research may be required.)

- Use your butter. Use it all. Ask for more. It's intended for the roll, but you can add it to rice, pasta, meat, or all of the above.

- If the meal includes a wedge of cheese, put it inside the roll to form a sandwich, then add butter and entrée sauce as condiments.

- If you're having breakfast on an international flight, know that few foods suffer more from precooking and reheating than eggs. They look and taste like the only food left for humanity in a dystopian future. Stick with cereal, fruit, or muffins.

- Experiment and take risks. After all, by putting any of that food in your mouth, you've already taken a risk. Shrewd investors aren't afraid to double down.

IN-FLIGHT SNACK AND BEVERAGE STRATEGIES

Get the most out of in-flight snacks and beverages:

- Whether ordering soda or juice, always (politely) ask for the whole can. You may even ask for a second, unopened can to save for later, thus creating what economists call a surplus. Or get a can of juice and a can of sparkling water and mix the two.

- Request extra pretzels, peanuts, and/or snack mix to help build your surplus. If a meal is served later, you can crumble the snacks over the food to improve the flavor and add crunch.

- Mixed drinks are one of the few good deals in air travel, because those little liquor bottles contain about one and a half drinks' worth of alcohol. Be sure to get a full can of whatever mixer you choose, plus an extra cup of ice. This allows you to regulate ratios as the ice melts and the effects of the alcohol set in.

IN THE EVENT OF BLANDNESS, YOUR CARRY-ON CAN BE USED AS A SAUCETATION DEVICE

A study by Lufthansa determined that when you're on a plane, your palate is up to 30 percent less sensitive to salty and sweet flavors, largely because the cabin is so dry. Airlines add extra seasoning to counteract this phenomenon, so meals taste better in flight. If you're packing food for a trip, you should do the same.

But you may still find yourself trapped at 35,000 feet with a bland-tasting meal. How best to avoid this cruel fate? Before your voyage, buy some empty, plastic, three-ounce travel bottles meant for shampoo and conditioner. Fill them with soy sauce, honey, curry, sriracha, or whatever potent potables float your boat. Pack them in your carry-on. When dinner time comes you'll be the envy of the aisle.

SMUGGLING ALCOHOL INTO CONCERTS AND SPORTING EVENTS

Like airports, concert and sporting venues take advantage of their captive audiences by charging outlandish prices for food and beverages. Shrewd Eaters combat these robber barons by sneaking their own alcoholic beverages inside.

Liquor is best, of course, because it packs the most punch and takes up the least space. Use a soft flask, like those favored by hikers, or the small plastic bottles of booze you get on an airplane.

Smuggling Alcohol in the Shoe

This is uncomfortable, but it doesn't need to last long. Put the bottle in your sneaker when you get close to the gate. Once you get inside go straight to the bathroom to remove it. They never check your shoes.

Smuggling Alcohol in the Jacket

This one's riskier but less painful. Security may pat down the jacket but they'll almost never make you open the hand holding the jacket.

RATIONAL EATING, COST-BENEFIT ANALYSIS, AND DIMINISHING RETURNS

The most fundamental question of economics as it pertains to our work is: Are Eaters rational actors? Do we always seek maximum deliciousness at minimum cost? Or do we sometimes act against our own gastroeconomic best interests?

Only someone who has never actually interacted with other human beings could think that we're all completely rational. We Eaters may understand objectively that we should try to get the most pleasurable experience for the least money, but outside influences and our own imperfections cloud our judgment.

How else to explain a bubble, in which Eaters pay more for something than it's worth simply because everyone else wants it?

We aren't as rational as we think we are, or as we want to be. But we can learn to reduce mistakes. I'll show you how to separate the signal of Perfect Deliciousness from the noise of the Eatscape in order to invest your money and stomach space wisely.

HOW RESTAURANTS WORK AND HOW YOU CAN WIN THEM

I spent three years waiting tables as part of my fieldwork for this book, so I can tell you that one word is drummed into servers' heads above all others: "upsell." That means no matter how much the people at your table are spending, you want them to spend more. If they order entrées, get them to order appetizers too. If they order soda, get them to put some liquor in it.

The most effective way to do this is with suggestive selling. Never simply say, "Want something to drink?" or "How about an appetizer?" Suggest specific menu items—put the picture in their heads, so they imagine having it in front of them.

Of course, you suggest different items to different customers depending on what you think they want, based on having sized up every ounce of their being within the first ten seconds of knowing them.

For instance, when I was a server at a pizza chain, I approached a young couple to take their order and noticed they were speaking Spanish. After they ordered their pizza, we had the following exchange:

ME: Would you like something to start with, maybe a salad?

THEM: [*blank stares*]

ME [*flustered*]: Uh, or, uh . . . nachos? How about some nachos?

WOMAN [*to man*]: *¿Quieres nachos?*

MAN: *Sí.*

That's an extra $7.69 on that check, baby.

When I worked at an upscale seafood chain, I met a server whom I'll call Kenny. He was a masterful salesman who could hold a table rapt with his description of a wine's tasting notes or a garlic butter's effect on a particular cut of a particular fish. He was making most of it up as he went along, but he sure was convincing.

Each night he picked a different expensive wine at random and told his tables it was the restaurant's "featured bottle" that night, even though the restaurant didn't actually feature any bottle on any night. Kenny could talk anyone into anything, which is why I did not find it surprising to hear several years later that he had inadvertently impregnated someone.

I recall being in the kitchen one evening when Kenny came bounding in, a large grin on his face. I asked him why he was so happy and he told me, "I just upsold my table to a really expensive bottle of wine."

"What did you tell them?" I asked.

He summarized the exchange like this: "They said, 'We'd like this cheap bottle of wine,' and I said, 'Why don't you get this expensive bottle of wine instead?' And they said, 'Why?' And I said, 'Bullshit, bullshit, bullshit.' And they said, 'Okay.'"

TURNING THE TABLES AT THE TABLE

All this doesn't mean you should ignore *everything* servers tell you. After all, while they want to make money, they also want you to be happy, so you'll tip well and come back. Understand the factors at play so you can interpret server cues and make profitable decisions.

- DO order something that the server suggestively sells you, if their pitch has reminded you that you truly wanted that item. Take care, however, to ensure that you don't get talked into ordering something you didn't actually want. That would be irrational.

- DON'T ask, "What's good here?" The server will be unsure of what you're after and is likely to throw out a bunch of options, skewed toward the expensive end of the menu, hoping that something strikes your fancy.

- DON'T ask, "Is X dish good?" What's the server going to say? No? Even if they think that dish is lousy they're not going to tell you, because for all they know, it's the only thing on the menu that looks appealing to you. If they cross that off your list, you're both screwed.

- DO ask for the server's input, while also offering some guidance. Say things like, "I'm in the mood for shellfish. What do you recommend?" Or narrow your options down to two or three and find out which the server prefers. This way you're helping them understand what interests you and eliminating dishes that you deem too expensive.

- DO follow up with more questions to find out exactly why they think one dish is better. The goal is to find out what *you* will like, not what your server likes. Details may help guide your decision and determine the authenticity of the recommendation. Unless Kenny is your waiter. Then you should probably ignore everything he says, or you could end up pregnant.

FINDING VALUE IN A CRAFT BEER BUBBLE

In recent years, many new craft brewers and small-batch distillers have brought increased quality and variety to the worlds of beer and spirits. But some proponents of this trend are growing bloated with self-regard, becoming as insufferable as the wine enthusiasts they once derided.

"Have you tried Beaver Snatch Brewery's Unabomber Porter?" they ask. "It's fermented in the same abandoned bus in Alaska where that kid from *Into the Wild* died."

Indeed, there are few more objectionable varieties of hipster than the craft beer enthusiast who claims the mantle of the commoner while displaying the affectation of the snob.

TIP Deliciousness can't be broken down into columns on a spreadsheet. It's no accident that even as investors increasingly rely on big data, the business world continues to revere swashbuckling CEOs who shoot from the hip and decide from the gut. If you want to satisfy your stomach, sometimes you must let *it* make decisions for *you*.

EXPLOITING THE DELICIOUSNESS OF LOSS LEADERS

A loss leader is an item that a store sells at a very low price, often at a loss, to lead customers inside, the idea being that once you're there, you'll buy other products and they'll make money off you in the end. That means that when a restaurant offers a special that seems too good to be true, sometimes it isn't. The rational Eater knows the difference.

When I lived in Chicago there was a bar that offered all the beer you could drink—and free snacks—for five dollars, every Friday from five to eight P.M. At eight their bar was full of customers, some of whom were still conscious enough to buy more drinks at full price. But I won by arriving right at five and eating a meal's worth of snacks, then building a beer reserve at seven thirty. This allowed me to drink for an extra hour and call it a night, all for only five bucks.

Loss leaders can be found at brunch too, where eateries offer a free mimosa or Bloody Mary with the purchase of a meal. It helps them sell more meals, and they know that if they can get you started on the sauce, there's a good chance you'll order a second or even third. And when you do, the restaurant makes its money back and then some.

If you really want another drink at brunch, order it and enjoy it. But if you want to get the best deal, drink your free drink and go home.

These people drive up the cost of good brews, encouraging a glut at the tap and rendering the term "craft beer" so overused as to border on meaningless. The market will eventually burst this beer bubble, but I fear it may forever tar my beloved beverage with a pretentiousness it had long avoided.

Not everyone who likes craft beer or small-batch liquor is evil—only a small but dangerous subset is. There are many great artisanal alcohols on the market, and some drinks are demonstrably better than others. But hype and higher prices don't always correlate with greater enjoyment.

My advice: Find a reasonably priced beverage that tastes good to you and enjoy it, secure in the knowledge that those paying more are, in economic parlance, "suckers."

ON MARKETING GIMMICKS AND TRUE EATOVATION

In recent years we've seen companies spend millions of dollars to subject us to ridiculous non-innovations, when a much simpler approach would have been far preferable.

How many different ways can pizza chains reconstitute bread, sauce, and cheese? Now the cheese is inside the bread! Now the bread is folded over the cheese! Now the sauce is on the side! Soon they'll just give you the ingredients individually and tell you to make it yourself.

Pizza purveyors, however, have nothing on the nation's largest beer brewers.

Coors Light has released several iterations of beer cans that change colors depending on how cold the beer inside is. In other words, they spent millions of dollars creating and marketing a device designed to do something that's already been done effectively for millennia by HANDS.

Miller Lite introduced the "vortex bottle," which has swirling grooves around the inside of the bottle's neck, meant to make the bottle and the pouring process look somehow more pleasing. (They didn't even pretend this would improve the beer itself.) But who cares how cool beer *looks* as it's being poured? It appears Miller confused alcoholics with stoners.

> **TIP** Bubbles from a carbonated beverage make you feel fuller than you are. When these bubbles burst, you realize that the gastronomic nest egg you thought you'd accumulated is gone. Take this deflation into account when investing your stomach space in beer and soda.

Bud Light has mostly avoided pointless alterations to their vessels, but that doesn't mean they haven't wasted plenty of money. They ran an entire ad campaign touting the beer's "drinkability." The slogan might as well have been, "Our product satisfies the most basic criteria of the definition of a beverage."

Rather than spend money developing asinine gimmicks, I'd rather these brewers donate a portion of their beer to the homeless, or simply tape a dollar bill to each six-pack. But if they insist upon trying to position themselves as thought leaders, they ought to at least pay their millions to a real innovator—me. If they do, here's what I'll invent:

- A can that stays cold even after you take it out of the fridge

- A bottle neck with interior grooves designed to modulate fizz, offering the pourer greater margin for error in head formation

- A pint glass that alters a beer's bitterness based on the weather

Beer industry giants, I await your call.

FAD DIETS: PONZI SCHEMES OF THE EATSCAPE

Everyone wants an investment with no risk and huge returns, just like everyone wants a diet that lets you lose weight and still eat whatever you want. People love to tell friends about their great new diet and enlist those friends to join the club. The truth for most people, however, is that successful dieting, like successful investing, involves incremental progress over an extended period, with occasional setbacks.

I know that doesn't sound very glamorous—I'm sorry I don't have better news. (Considering how much money diet books make, I suspect my publisher is sorry too.)

People who throw their money at each new diet craze may start off winning big at the buffet table, but they end up losing everything, by which I mean regaining everything. Eventually they come to realize that they've invested vast stomach resources and received nothing but diminished pleasure in return.

Consider the Paleo Diet. Only irrational actors could be convinced that in order to live healthier lives, we should eat just like they ate when the average life expectancy was sixteen. I'm all for natural eating, but there is such a thing as progress. What's next? Will trendy gyms get you to run faster on the treadmill by putting a saber-toothed tiger behind you?

TIP The first few bites of anything are almost always the best ones. After that, it's diminishing returns. One exception is truly great steak, which makes me hungrier with every bite. I call this phenomenon the Steak Satiety Paradox. Technically speaking it should mean that you can eat great steak forever, but I have yet to determine whether that's the case. I always run out of steak first.

I've even seen mention of such things as "paleo comfort foods." Come on. There was nothing comfortable about living in the Paleolithic Era. A good day back then was one in which you *weren't eaten*. A bad day meant you were halfway down a cave lion's digestive tract, in which case you weren't going to get much comfort from mama's yakloaf or a nice hot bowl of gruel.

EXPENSIVE FOODS THAT ARE WORTH IT

- Farm-fresh, local eggs, from well-treated chickens
- A really good lobster roll, on a top-sliced, buttered, griddled bun, with chunks of lobster that are large and plentiful
- Great espresso

CHEAP FOODS THAT AREN'T WORTH IT

- Potato chips that aren't kettle cooked
- Margarine
- Grapefruit

CLASSICAL VS. KEYNESIAN ECONOMICS

Classical economists, inspired by Adam Smith, argue that government intervention gets in the way of growth and that free markets regulate themselves. Smith says that in the process of furthering our own self-interest, we are "led by an invisible hand to promote an end which was not part of [our] intention," one that benefits society as a whole.

Such is the case when you overestimate your stomach's demands, make a huge tuna casserole, and realize you couldn't possibly eat the entire thing yourself. You share it with your family, making clear that the next time one of them is cooking something, you get a piece. These mutually beneficial familial transactions, and the tenuous interpersonal relationships they engender, are but one of the many blessings of a free-market kitchen.

But there's another field of thought. The Keynesians, led by John Maynard Keynes, argue that markets left alone will run amok, so regulation is vital. In other words, extra casserole doesn't *always* find its way to hungry mouths as a natural by-product of self-interest. In fact, an unchecked free market encourages dangerous swings in casserole supply and demand. When its value plummets and there's a bunch of casserole nobody wants, the trays and trays of leftovers are too big to foil. Keynes says society must intervene to stop things from getting that bad.

In truth, each of these approaches has something to teach us, both on the micro level, in terms of regulation of your own stomach, and on the macro level, in terms of the foods you put into it. In a sense, you are the Fed chairperson of your own personal eatconomy. It's up to you to set policies for healthy growth, by understanding when to intervene and when to leave matters to the market.

ON REGULATION OF YOUR STOMACH EATCONOMY

If your stomach were a bastion of pure classical economic theory, it would be free to consume as much as it wants, whenever it wants. This might lead to overconsumption and inflation of the waistline, in which case the invisible hand would respond accordingly, perhaps with a lethal blow to the lower intestine.

In order to avoid such extremes, a dose of Keynesian intervention is required. Here are some of the tools at your disposal, and their intended effects:

- **LOOSEN YOUR BELT** When high-quality foods are at hand, you want to create the conditions for stomach growth. Loosening your belt is the equivalent of lowering interest rates, and encourages you to spend more gastroeconomic resources.

- **TIGHTEN YOUR BELT** Like raising interest rates, this encourages conservation. When you want to temporarily restrict the stomach while hungrily awaiting a good meal, this is your move.

- **TAP STRATEGIC RESERVES** The reserve requirement is the amount of stomach space that you must always keep available in case of emergency. You should only tap into this resource when especially delicious opportunities arise.

- **USE QUANTITATIVE APPETITE EASING** You may have noticed that when you're very hungry for a very long time, you actually lose the desire to eat. It's as if your taste buds have been unemployed for so long, they've stopped looking for work. If you feel this happening, eat a little bit of something— anything—to keep your appetite up. The food will taste great, no matter what it is. This type of artificial injection of deliciousness can provide the stimulus your stomach needs to make it to more profitable times.

TOXIC ASSETS CAKE

If your fridge has gone an especially long time without intervention, you may have some cake in there that's so spoiled, Keynes himself couldn't save it. This recipe is perfect for pawning it off on unsuspecting neighbors. The key is to combine tranches—French for "slices"—of many different cakes, some of which may be more palatable than others.

YOU WILL NEED:

6 tranches of quality cake

3 tranches of decent cake

3 tranches of rancid cake

INSTRUCTIONS

Place all 12 tranches in a blender and mix until the different cakes are indistinguishable, then press them into a cake pan and cover with a fresh layer of frosting. Serve to second-tier friends, those you can afford to lose.

One great thing about this dish is that it usually takes a few days before it makes people sick, so they won't be able to trace the food poisoning back to you. If you're really lucky, they'll regift the cake to someone you don't even know, and you'll be beyond the reach of the invisible hand.

Note: DO NOT EAT TOXIC ASSETS CAKE.

CEREAL AND MILK: WHERE CLASSICAL MEETS KEYNESIAN

Cereal itself epitomizes the blessings of the free market. Its many shapes, sizes, colors, textures, densities, and flavors could only be possible in a land that encourages innovation, consumption, and the consumption of innovation. (You think Chairman Mao got to pick from twelve varieties of Cheerios?)

The free-market approach also guides us when combining different cereals together, one of the Eatscape's great simple pleasures and an excellent entry point for youngsters. There are really no rules—regulation here only interferes with risk-taking and new ideas. The merits of any combination are simply a matter of personal preference. Each will sink or swim in your stomach economy based on its level of deliciousness.

However, adding the milk to the cereal is a whole different story. That process is a calculated intervention that requires regulation and strict adherence to proscribed techniques. In other words, it calls for the strong hand of John Maynard Keynes.

Let's delve into both cereal mixing and milk application to learn more about this confluence of classical and Keynesian economics.

COMBINING CEREALS: THE FREE MARKET AT WORK

It's up to you and the market to determine which combinations of cereals are best. But there are two recurring themes from this book that you'll want to keep in mind before you begin. (Even the free market is subject to science.)

Surface-area-to-volume ratio (SATVOR) is crucial, as it helps dictate the rate of soggability. (The physical composition of the cereal itself is also key, of course.) A higher SATVOR generally means milk will be absorbed faster and crunch will be shorter-lived. Flakes turn soggy faster than O's, which turn soggy faster than solid bits without holes. Nuts are like fortresses.

This is not to say that all the cereals you mix together must have the same SATVOR—textural variety can be a strength. You may find you enjoy a combination that, after sitting in milk for a few minutes, includes crunchier and soggier components in the same bite. It's just something to take into account.

The other question you must answer is, do you want bite consistency or bite variety? If bite consistency, mix the cereals before adding milk, possibly using the Claw method (from the snack mix section on page 251 of "Biology and Ecology"). Just note the Brazil Nut Effect, also known as the

Muesli Effect, which states that when you mix foods of different weights, the heavier ones eventually end up on top.

If you want bite variety, leave the cereal in the layers in which you poured them into the bowl, keeping in mind that the one you pour first will be at the bottom, so it will sit longer in milk.

Beyond that, just remember that there's always room for more cereal innovation. Don't let the free market's rejection of Bill and Ted's Excellent Cereal dissuade you. Act in your own self-interest, pursue deliciousness as you see fit, and we'll all benefit. Try the recipes below for truly excellent results, then test your own concoctions.

DESSERT CEREAL COMBOS

These are great desserts, whether you're curled up on the couch watching TV or looking for the perfect conclusion to your dinner party.

Count Chocula or Chocolate Lucky Charms
+ Golden Grahams = S'mores

Apple Jacks + Cinnamon Toast Crunch
+ Cracklin' Oat Bran = Apple Crisp

Frosted Mini-Wheats + Honey Bunches of Oats
+ nutty granola = Baklava

For another innovative cereal recipe, see "Cheddar Goldfish Cereal" on page 100 in "Philosophy."

MOCHA MORNING CEREAL

There's tremendous potential for new cereal concepts when you tinker with the liquid component. To make Mocha Morning, combine equal parts milk and coffee and pour over any chocolatey cereal.

CHEERIO CEREAL

This is similar to Mocha Morning but with an English vibe. Combine equal parts milk and tea and pour over any honey-based cereal, such as Honey Nut Cheerios.

CEREAL COMBOS AS BREAKFASTS OR DESSERTS

I stand by my statement that there is no objectively wrong cereal combination, but I can't resist offering my own opinion. When I mix up a bowl of cereal, it's usually for either breakfast or dessert, and that distinction guides me. Because I shy away from large quantities of sugar in the morning, my breakfast combos generally look like this:

- **50 PERCENT BASE CEREAL**—something healthy and simple, in the middle of the flavor and crunch spectrums
- **25 PERCENT SUGAR CEREAL**—something sweet and bursting with deliciousness
- **25 PERCENT CRUNCHY AND FILLING**—something with protein and fat (but little sugar), along the lines of Grape-Nuts, a low-sugar granola, or straight-up nuts

When evening falls, however, my cereal combinations take on a very different tone. See "Dessert Cereal Combos," page 149, for details.

THE KEYNESIAN SCHOOL AND MILK REGULATION

Cereal innovation would not be possible without Adam Smith's classical free-market approach. But you don't want the invisible hand adding your milk. Pouring techniques must be regulated, or you risk severe swings in milk supply and demand, even within a single bowl. Your approach should take one of two forms—the Drizzler or the Single Stream (figures 5.5 and 5.6).

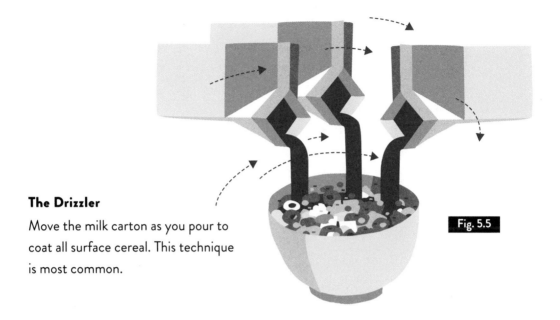

The Drizzler

Move the milk carton as you pour to coat all surface cereal. This technique is most common.

Fig. 5.5

The Single Stream

Sometimes a different type of milk intervention is required, especially with high-SATVOR cereals that turn soggy quickly. With this approach you pour all the milk in at a single spot, so as much surface cereal as possible stays dry.

Fig. 5.6

Your consumption strategy can also help prolong crunch. Instead of eating the entire top layer of cereal first, start in one quadrant and eat down to the bottom of the bowl before moving to the next quadrant. This way some surface cereal stays dry and crunchy until you eat it.

WHEN TO ADD THE MILK BEFORE THE CEREAL

No doubt you've had the misfortune of filling a bowl with a heaping pile of your favorite cereal, only to realize too late that you're low on milk. When you tip the carton, a single, sad splash of the white stuff plops out. At that point, most of the cereal in the bowl is wet, so you can't put it back in the box. But you're out of milk, which means your optimal ratio is unattainable.

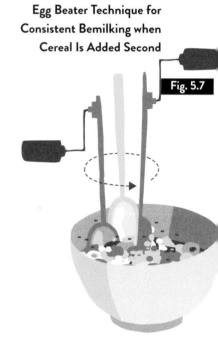

Egg Beater Technique for Consistent Bemilking when Cereal Is Added Second

Fig. 5.7

The solution is to check milk levels before doing anything else. When it's in short supply, pour it into the bowl first. This way you can add just the right amount of cereal for that amount of milk. You lose the effect of pouring the milk over the cereal to wet it, but maintaining your ideal cereal-to-milk ratio is paramount. In this case, you can replicate the effects of more traditional pouring with the Egg Beater Technique (figure 5.7). This technique makes the best of an already bad situation.

HOMEWORK

It's a dog-eat-dog world, and if you want your palate to turn a profit, you need to know how to compete in the modern Eatscape economy. The principles in this chapter have prepared you to do just that. Here's your assignment . . .

Explain the positive correlation between national economic distress and local yogurt varietal popularity, as seen in recent years in Greece and Iceland.

Submit your homework to me at dan@sporkful.com.

6

CULTURAL STUDIES & ANTHROPOLOGY
Holidays and Everydays

- Thanksgiving advice from competitive eaters and noble savages

- The Veggieducken

- The cruel deceptions of the Easter Bunny

- Mac and cheese analyzed, crispified, and goofied

- The genius of spiral-cut ham and new uses for spiral-cut technology

- Adding pork to classic Jewish foods (the Heretic's Buffet)

- Passover Sangria and Manischewitz Sorbet

- Eating at work, including cube farm consumption techniques

- Vehicular eating and drive-thru strategies

There are many different ways to measure our progress toward Perfect Deliciousness, but the most tangible signals of change come from the culture that surrounds us.

As the anthropologist Margaret Mead reportedly said, "Never doubt that a small group of thoughtful, committed citizens can change the world; indeed, it's the only thing that ever has." If we're going to revolutionize the Eatscape, we must study our culture's social norms and customs, so that we can create and spread some of our own.

I'll divide our discussion of eating culture into two categories: holidays and everydays. Holidays have the most ritual and folklore associated with them, and the food is often part of the fanfare. So we'll discuss the importance of a momentous Thanksgiving, the wonders of a spiral-cut Christmas ham, and the problems with the Easter Bunny.

Then we'll transition to more quotidian culture and cover two mainstays of the everyday—eating in the car and eating at work. How best to set up your fry-dipping station after leaving the drive-thru? How can the office vending machine be used most effectively? And how does the nature of the workplace gathering dictate the way in which the sheet cake should be sliced?

Even the ordinary is rife with ritual, which means opportunities to increase deliciousness abound.

THANKSGIVING AND NEW YEAR'S EVE

Thanksgiving is the best holiday of the year and New Year's Eve is the worst. I want to address them together to show you that even a great eating experience can be improved, and even an awful one can be saved.

THE NOBLE SAVAGE'S APPROACH TO THANKSGIVING DINNER

Anthropologists like to travel to remote villages to study people and societies cut off from the rest of the world. In part it's to try to get an idea of how our ancient ancestors lived, and in part it's to see what these villagers can teach us.

As it turns out, it's quite a lot, because Thanksgiving dinner requires both strategic precision and barbarism. Let us look to the timeless folk wisdom of the Eatscape for practical advice on turkey day:

- **"DON'T HUNT A DOG WHEN AN ANTELOPE IS NEAR."** Focus on the relative availability of each food. If you can get it or make it yourself without difficulty throughout the year, skip it. Special occasions are for special foods.

   Turkey is a prime example. We don't cook whole turkeys very often because it takes a long time. Congress declared Thanksgiving a national holiday to give Eaters one day to devote solely to the task. (Note to vegetarians: You too can enjoy a time-consuming and elaborate dish fit for special occasions like Thanksgiving. See "Behold the Veggieducken" on page 160.)

- **"HE WHO BRINGS BUT A CUP TO THE RIVER IS SOON THIRSTY AGAIN."** Prepare for stomach expansion by wearing loose-fitting pants, or a garbage bag with a drawstring top and leg holes.

- **"THE SNAKE WHO STRIKES EARLY WILL ALSO STRIKE LATE."**
  Push for an early start to the meal so you can recover in time for an evening feeding.

Certain principles of buffet strategy apply here as well. Survey all the options before taking anything. If a promising new dish is present, take very little at first, then go back for more if it's worthy. Avoid fillers like bread, salad, and soda.

A THANKSGIVING PEP TALK FROM A COMPETITIVE EATER

At the time of this writing, Tim "Eater X" Janus is the second-ranked competitive eater in the world. That means he can eat more food in less time than about 5,999,999,998 of us. He's known on the competitive eating circuit for his fearsome face paint and for once showing up at the Nathan's Hot Dog Eating Contest holding a sign that read, SPOILER ALERT: HERMIONE DIES.

When Eater X came on *The Sporkful*, I asked him to offer some words of wisdom to Eaters about to sit down to Thanksgiving dinner. Eating your best doesn't mean eating until you're physically ill, and even Eater X doesn't approach every meal this way. But on Thanksgiving, we're all competitive eaters.

Here's what he said:

> It comes down to mental strength and discipline. If you're going to do something, you want to do it as well as you can, so you don't regret not giving your best effort. Once you set your goal, if you quit, you're going to have to wait a whole year to try again, and you're going to be angry at yourself for a whole year.
>
> When I compete, I know there's going to come a time when I want to quit, and I want to have it in my head that I won't allow myself to quit. You can go a lot further than you think you can go. Stay strong.

Powerful words from a man who once ate two gallons of chili in six minutes.

ON TURKEY COOKERY

The preparation and consumption of the turkey is a special occasion, a beloved tradition to be savored. Don't rush it.

The more elaborate your turkey preparation rituals, the more cultural folklore will develop around it and the more delicious the turkey will taste. That's why I recommend brining your bird. It seasons and softens the meat, and requires you to start turkey preparation at least a day in advance. Your extra effort will imbue the feast with the unmistakable flavor of momentousness.

The Internet can tell you the basics of how to cook the bird, but here are some techniques to add both deliciousness and import:

- Stuff the turkey. (See also "If You Prefer Dressing the Turkey, Stuff It" on page 159.)

- Cook the turkey upside down on a rack for the first hour and a half to two hours. Cover the rack in foil, poke holes in the foil, and coat the turkey in olive oil. This protects the breast from the oven's heat, keeping its internal temperature down and conserving moisture. Flip the bird using oven mitts. The breast may not look as pretty, but you don't eat with your eyes, and all the white-meat suckers will be so thrilled with the results that they'll keep their hands off your dark meat.

- Some people argue against basting, but I believe it adds vital grandiosity. And if nothing else, it gives you something to do while you're drinking.

- While cooking the turkey, drink.

- Never wait for that plastic needle to pop up. Those things were designed by lawyers who want your turkey so overdone that you can't possibly sue them for giving you food poisoning. Cook the bird until the temperature is 160 degrees in the thickest part of the thigh. Make sure your thermometer isn't touching bone. Take a couple of readings to be sure. After removing it from the oven, let it stand for thirty minutes before carving.

TASTE THE MOMENTOUSNESS TURKEY BRINE

This brine recipe is for a 14- to 18-pound turkey. If your turkey is larger, add more water and brine it for longer—up to 36 hours.

YOU WILL NEED:

water

1 cup kosher salt

1 cup teriyaki sauce

1 cup low-sodium soy sauce

1 cup garlic powder

1 cup onion powder

> **TIP** As you plunge your turkey into the brine, state loudly and with authority, "Initiate brining sequence!" It feels really awesome.

INSTRUCTIONS

Make sure your turkey is fully defrosted. In a 5-gallon bucket, mix 3 gallons of cold water and all brine ingredients. (You can mix the brine in a pot and then use a brining bag, but those things scare me. They're liable to open, and it's hard to store them in the fridge in such a way that the sides don't sag, leaving the turkey unsubmerged.) Stir until salt and spices are dissolved. Add several handfuls of ice.

Place the turkey—innards removed, breast side down—in the brine so it's completely submerged. Add a little more cold water or ice if necessary. Cover and put in the fridge.

You want the bird in the brine 20 to 24 hours before it goes in the oven.

On Thanksgiving, when you're ready to cook, pour yourself a drink. (Remember, this is an event.) Preheat oven, remove turkey from fridge, discard brine, wash bird in cold water, and pat it dry with paper towels. Cook it as you like.

IF YOU PREFER DRESSING THE TURKEY, STUFF IT

The terms "stuffing" and "dressing" are often confused and conflated. Stuffing is stuffed inside of the bird and dressing is used to dress the bird. (Animals, like people, wear their clothes on the outside.)

Let it further be known that stuffing is superior to dressing in every way.

1. Stuffing absorbs glorious meat juices and flavors from the animal in which it's stuffed. Dressing does not.

2. Stuffing imparts glorious flavors to the animal in which it's stuffed. Dressing does not.

It's true that even a large turkey can't hold enough stuffing for a big Thanksgiving feast. So make a separate pot of dressing and mix it with the stuffing from inside the bird. But you should still call it stuffing.

TIP Some people complain that a stuffed turkey takes longer to cook, but if your goal is to get it over with, you are ignoring a vital component of this ritual observance.

THE STUFFING SANDWICH: A BREADLESS VESSEL FOR THANKSGIVING LEFTOVERS

The Thanksgiving leftovers sandwich is a classic ritual, but one that I believe is in need of improvement. Stuffing is mostly bread, and mashed potatoes are another starch. A roll just adds a lesser version of the same type of food, and gets in the way of the featured attractions. That's why I've created a Thanksgiving leftovers sandwich without traditional sandwich bread.

Put 2 baseball-sized balls of stuffing in a bowl. Beat 1 egg and mix it thoroughly into the stuffing. Form the stuffing back into 2 balls and place them into an amply-oiled, non-stick pan. Press them into patties about ¼ to ⅓ inch thick and fry until dark golden brown and crispy on the bottom. (The oil should come roughly halfway up the sides of the patties.) Flip and fry some more.

When they're crispy all around, remove them to a rack to cool a bit before using them in place of sandwich bread. Note that stuffing with large chunks may not hold together well enough for this to work.

MAXIMIZING LEFTOVERS WITH NEW THANKSGIVING TRADITIONS

Thanksgiving leftovers are a wonderful part of the holiday observance, but a few days after the big day, you may get tired of eating the same food and/or stressed about its going bad before you've finished it. Here are some tips:

- **DON'T COOK—RECONSTITUTE** Don't make new food to go with the old food. Turn leftovers into new foods. Make turkey soup, turkey hash, a stuffing sandwich (see recipe), or your own creation.

- **USE BREAKFAST** Candied yams are a great way to start the day, provided they aren't too sweet. Spread cranberry sauce on toast in lieu of preserves.

- **THROW A LEFTOVERS EXCHANGE PARTY** A few days after Thanksgiving, you're likely sick of two things—your family and your leftovers. So invite all your friends and their leftovers. Lay all the containers on a table, put out a pile of paper plates, and eat up. This way you can get rid of excess food, eat something different, and learn new recipes for next year.

BEHOLD THE VEGGIEDUCKEN

Vegetarians don't like it when you suggest that they're missing out by not eating turkey at Thanksgiving. But while there are many wonderful vegetarian dishes to be enjoyed, the truth is, they *are* missing out.

That's not because you need to eat meat to celebrate Thanksgiving. It's because you need to cook something time-consuming and huge. Vegetarians have never had that one large, elaborate, centerpiece dish that befits a major holiday.

That's why I invented the Veggieducken.

THE VEGGIEDUCKEN

This dish made its debut on my Cooking Channel web series *Good to Know*. It's a takeoff on the Turducken—a chicken inside a duck inside a turkey, with stuffing between the layers—made of two sweet potatoes inside leeks inside a banana squash, with vegetarian stuffing between the layers.

Vegan Eaters Mirjam Lablans and Dan Goldstein of Brooklyn were inspired to try making a Veggieducken and said it accomplished its goal.

"It was a lot of fun," reports Dan. "It definitely gave me a sense of having been more involved with the process of creating the Thanksgiving centerpiece than I normally get to experience as a vegetarian."

One important tip: Banana squash can be hard to find in some areas, so you can use two butternut squashes instead, putting one sweet potato inside each. You'll slice the whole thing up before serving anyway, so you can still put it all on one big platter.

As you build from the bottom up, it's half a squash (seeds removed), stuffing, leeks, stuffing, sweet potatoes microwaved for 3 to 4 minutes before insertion, stuffing, leeks, stuffing, other half of squash. Bake at 350° for 2 to 3 hours until it's soft throughout. Slice and serve.

Veggieducken Illustrated

ON THE PROBLEM WITH DEEP-FRIED TURKEY

Frying a whole turkey sounds exciting, and any time there's a decent chance your dinner could burn down your entire house, that's always a plus. But deep-fried turkey is overrated.

To be clear, I don't object to its being fried. Rather, I object to its not being fried *enough*.

What makes fried foods great? For the most part it's the taste and texture of the exterior—the part we Eaters call the "fried"—as well as the fried's ability to lock in the juice and flavor of the food it surrounds.

If you want a great fried food, you need a lot of exterior to cover with fried—a high SATVOR. However, a whole turkey has very little exterior in comparison to its large interior volume. When deep-fried it may come out juicy, but relatively few of the bites will contain fried.

If you want to cook your turkey this way, prepare it as you would a chicken: Cut it up. Bread and fry each drumstick, thigh, and wing individually. Cut each breast once the long way and three or four times the short way and fry each of those pieces. This will increase your amount of fried per bite exponentially. And if it burns down your house, at least you'll have a better excuse.

AVOIDING NEW YEAR'S LETDOWN SYNDROME

Thanksgiving is the best holiday because it's built for Eaters. The whole point is to partake in a bounty for which you give thanks. New Year's Eve is the worst because so many forces conspire against us.

Popular culture dictates that your New Year's must be the biggest, best party of the year—which is precisely why it never is. Restaurants tack a glass of champagne on to a mediocre prix fixe menu and expect you to mortgage your house to pay for it. Bars offer all-you-can-drink specials, then pack the place so tight that you can't get to the bar.

Want to enjoy New Year's Eve? Here are the keys:

1. Lower your expectations.

2. Stay home or go to a friend's. Don't go to a restaurant or bar.

3. Save yourself for New Year's Day.

New Year's Day is a great opportunity for a casual get-together or open house party, and a chance to taunt your friends who spent the previous night waiting for a drink that would never come.

MAKE-YOUR-OWN NEW YEAR'S FESTIVITIES

Most holidays come with a traditional way to celebrate them. There's the Thanksgiving turkey, the July 4th barbecue, and the St. Patrick's Day corned beef and cabbage. What sets New Year's Eve/Day apart culturally and anthropologically is that it's one holiday where some sort of festive gathering is the norm, but the particular eating rituals are less established. That's why, whether you focus on the Eve or the Day, it's the perfect holiday for make-your-own food events. Such as:

- Sundaes

- Pizzas

- Grilled cheese sandwiches

- Sushi (It's easier than you think. Google it.)

TIP All holidays are either casserole holidays or dip holidays. The difference lies in whether the most prominent layered food is eaten with utensils or chips. Thanksgiving and Christmas are casserole holidays. July 4th, New Year's, and Super Bowl Sunday are dip holidays.

Assemble a variety of base ingredients and let guests (including kids) get creative. Offer some obvious options and a few wild cards. Set it up so people can make small portions, which allows them to try several recipes before getting full. These types of parties are also more fun than a sit-down dinner because they get people moving around and offer an indoor group activity during a time of year when it's often too cold to be outside.

CHRISTMAS AND EASTER

I may be Jewish, but a lot of my best friends are Christian. Over the years they've become accustomed to my peering at them while scribbling furiously in my field notebook, which is why I feel qualified to address these holidays here.

ON THE PROMISE OF SPIRAL-CUT TECHNOLOGY

I lived with a Christian roommate for two years in order to study her species up close. At Easter, her family traveled to our abode from a remote village in northern Michigan and prepared various traditional dishes.

That was the first time I ever saw a spiral-cut ham.

I studied this bizarre creation, staring at it like a Sentinelese islander looking at Skype. The spiral-cut ham is a great technological advance for several reasons:

1. Portioning could not be easier, and you'd have to be an excellent carver to leave less meat on the bone.

2. It's much more convenient to snack on leftovers. Simply open the fridge and tear off a piece, as I did repeatedly as part of my research.

3. You could eat the entire ham in one bite—just pick it up and start chewing and twirling simultaneously. You know when cartoon characters put a huge piece of meat in their mouths, then pull out a clean bone? That could be you, thanks to the miracle of spiral-cut technology.

4. You could unfurl the entire thing and wrap it around a loved one.

5. The spiral is an inherently elegant shape and thus aesthetically pleasing to members of all cultures and tribes.

Clearly, more foods must be spiral cut.

The fine fried food purveyors of America's state and county fairs have already recognized this fact, which is why you may have seen fried spiral-cut potatoes on a stick right next to the fried butter. But this is just the beginning. Here are some other foods would benefit from spiral-cut technology:

Spiral-Cut Watermelon

Cut off the ends of the watermelon. Insert the handle of a long wooden spoon into the center of one end and push it out the other end. (You may feel a bit like you're taking the watermelon's temperature with a rectal thermometer.) Slice down to the spoon handle, using it as your guide, then rotate the knife and/or the watermelon around and around in a spiral. Keep the knife against the spoon handle. Continue your single spiral cut until you get to the other end. Before serving, gently pull on both ends of the watermelon to separate the spirals just a little, which creates a stunning visual effect. This lets you slice different-size wedges to order for different-size faces. In lieu of the wooden spoon, you can also use a vertical paper towel holder.

Spiral-Cut Spiral-Cut-Ham Sandwich

EAT THE EASTER BUNNY

Because the Easter Bunny has no real religious significance, I can shed the anthropologist's pretense of cultural sensitivity and lay bare this creature's cruel deceptions.

Most Easter candy is pretty bad, and gets worse when molded into the shape of a bunny. (Chocolatiers have discovered they can unload their worst product when they package it in mammalian form.)

Fried onions from the can are typically associated with the common Christmas and Easter dish green bean casserole, but they have so many other uses. Add them to salads, sandwiches, vegetables, wraps, pastas, rice, couscous, pizza, burgers, mac and cheese, soups, and grilled onions. And everything else.

But when you remove the shiny, colorful, chocolate bunny's wrapper, you find that the detailing work is terrible and the bunny is hollow. This is the first deception.

The second is the entire connection to eggs. Bunnies don't lay eggs. The Easter Bunny has no reason to have or protect eggs. Perhaps it's similar to a benevolent dragon who guards a mixed treasure procured under mysterious circumstances, then shares his trove with children. But I don't give the Easter Bunny that much credit because . . .

Bunnies are evil. This is the third deception. They look cute and cuddly but they're actually ornery and mean, and want nothing more than to bite you. (There's a reason why they love carrots—it's the vegetable most similar to the human finger.)

The only positive contribution that bunnies can make to Easter is with their lives, and only through this act can they repent for their sins.

So let us cease and desist all Easter Bunny iconography, chocolate and otherwise, and instead make a place at the Easter table for some delectable, tender rabbit. Once the children taste it, I'm sure they'll understand.

Ham Hock Mistletoe (or, Mistlehock)

I love the idea of designating an area of the home for holiday ardor, but I question the choice of mistletoe. Disembodied foliage is not especially romantic. That's why I think it's time for our culture to embrace an alternative that's as soft and succulent as a lover's kiss—a ham hock. Or as I call it, Mistlehock.

MAC AND CHEESE BEST PRACTICES: OVEN, STOVETOP, OR YES?

Holidays are a time for comfort foods, those dishes that make your stomach feel like it's sitting by a glowing hearth all wrapped in a cozy blanket. In many households, that means mac and cheese. But while the visceral solace it provides is undisputed, its ideal preparation is a matter of great debate.

Some believe mac and cheese is best cooked on a stovetop, while others favor the oven. The stovetop variety is usually cheesier and saucier, while the oven variety tends toward dryness. But the oven's arid heat allows for a crispy toasted topping—impossible on the stovetop.

Which is better? This age-old debate is a false choice.

The solution is to make mac and cheese on a stovetop in a saucepan, pour it into a baking dish, cover it with bread crumbs, and bake it in the oven for just a few minutes, until the topping is golden brown and crispy. It's the best of both worlds.

> **TIP** In determining the ideal macaroni shape for mac and cheese, sauceability is most important, as the pasta is primarily a cheese sauce delivery system. Forkability is least important, as the sauce helps hold pasta together and keep it on your fork. (See short pasta principles on page 260 in the "Genetics and Taxonomy" section of "Biology and Ecology," for a more nuanced discussion.)

THE CHEESE MUST STAND ALONE

What's the best cheese for mac and cheese? There is no single right answer. I recommend combining two varieties: one that's mild and creamy and one that brings what cheeseophiles call "a little something something."

Jack the Horse Tavern in Brooklyn Heights, which serves my favorite mac and cheese of all time, uses smoked gouda and fontina. American cheese is always a great option too, with or without something funky. Cheese is the real star of this dish, and any good combination of cheeses will be delicious on its own. If the cheese stands alone, it will stand well with the mac.

JACK THE HORSE MAC AND CHEESE

The folks at Jack the Horse restaurant in Brooklyn Heights make the best mac and cheese I've ever eaten. It has the perfect combination of cheese flavors, creaminess, and crisp, plus a ridged, corkscrew mac that maximizes cheesehesion. Chef/owner Tim Oltmans was nice enough to let me publish his recipe here. Enjoy!

Makes 4 servings

YOU WILL NEED

1 pint potato cream (see recipe below)

12 ounces grated cheese (8 ounces smoked gouda and 4 ounces fontina)

2½ cups cooked cavatappi pasta

1 tablespoon Dijon mustard

¼ cup toasted panko bread crumbs

pepper

INSTRUCTIONS

Gently heat potato cream. Add cheese; stir until melted. Add pasta and mustard; stir until heated. Fill baking dishes and sprinkle on bread crumbs. Bake at 400° for 6 minutes (if making mac and cheese ahead bake for 8 to 10 minutes when ready to serve).

Note: Individual baking dishes are best. You want dishes that are not too deep or they'll take too long to heat and the fat may separate.

POTATO CREAM

YOU WILL NEED

1 pint heavy cream

½ cup grated potato

¼ teaspoon grated nutmeg

½ teaspoon salt

½ teaspoon pepper

Heat all ingredients, stirring frequently until thickened, 15 minutes or so. Strain out and discard grated potato. You should have about one and two-thirds cups of liquid left, and it should be like béchamel (roux and milk). This process washes the starch off of the potato to thicken the cream and absorb the fat from the melted cheese.

MAC AND CHEESE AND ?

Just because you can put almost anything into mac and cheese doesn't mean you should. Mac and cheese is by its nature a decadent dish, rich and flavorful. It's not a place for mild and delicate additives that will get lost, or expensive foods that won't be fully appreciated.

Anyone who pays money for lobster mac and cheese is a rube. And if someone gives it to you for free, pick out the lobster, wipe off the cheese, and eat that first.

Furthermore, while smoky meats can be great additions to mac and cheese, *more* is not always *better*. Mac and cheese is beautiful in its simplicity.

If you do want a more elaborate experience, consider using the mac and cheese in a new and different way, rather than just dumping more food into it. Here are two examples of real mac and cheese innovation . . .

Chili Mac and Cheese Dog Bowl for Humans
The entire bowl is lined with mac and cheese before it's filled with chili and hot dogs.

LEFTOVER HAM SANDWICH WITH
MAC AND CHEESE SPREAD

Whether your Christmas ham was spiral cut or not, you'll love this Boxing Day treat. Scoop mac and cheese onto a cutting board and chop it up, then use it as a spread on a leftover ham sandwich. To put the focus on the mac, spread it thickly on both sides of the sandwich. To bring the ham to the fore, put the mac only on the top and use less of it.

HANUKKAH AND PASSOVER

I've studied Jews in all sorts of environments, from suburban New Jersey to New York City to a paved archipelago known as Long Island. Customs vary from one region to the next, but certain rituals are universal.

Even if you're not Jewish, I urge you to read this section. Some of the foods associated with Jewish holidays have great potential to better the larger Eatscape. Others could use some bettering themselves.

TO EVERY POTATO PANCAKE THERE IS A SEASON,
AND THAT SEASON IS ALWAYS

Traditionally eaten at Hanukkah, potato pancakes (or latkes) are essentially like McDonald's hash browns, except homemade and delicious. They're great not only because of their crispy fried potato grandeur, but also because their flat shape and firm structure mean you can put them in any sandwich, or use them in place of sandwich bread itself.

These delicacies remain a largely untapped resource in the Eatscape and should be eaten both more frequently and in more different ways. Here are some options:

- Serve with sunny-side-up eggs and dip them into the yolk.

- Serve alongside saucy foods and stews so they can soak up some liquid while maintaining their crisp.

- Insert into sandwiches of the meat, cheese, veggie, or egg variety to add crispy fried potato goodness.

- Use in place of sandwich bread for virtually any sandwich. (Ever had lox and cream cheese on a potato pancake? If not, go do that right now and come back when you're done.)

- Employ as the base for crostini, as in the Heretic's Buffet on page 172.

THE HANUKKAH MIRACLE SANDWICH

Hanukkah celebrates a miracle at the ancient temple when the Jews thought they only had enough oil to light the candles for one night, but the oil lasted eight miraculous nights. That's why foods cooked in oil are a common part of the Hanukkah observance.

American Jews eat fried potato pancakes, but in Israel, Jews celebrate with a different oily, fried food—donuts. I've brought these two rituals together to create a new sandwich: the Hanukkah Miracle.

It combines sweet and savory, sugary and tart, doughy and crispy. You may also make an Inside Out Hanukkah Miracle, putting the donut in the middle and potato pancakes on the outside. You'll just need pancakes that are relatively large in diameter.

Donut interior is griddled in butter and flipped inside out.

Potato pancakes are hot and crispy.

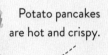

Sour cream and applesauce are at the Temperature of Life. Sour cream is on the bottom. Serve with extra sour cream on the side for dipping.

ALTERNATE USES FOR HANUKKAH GELT

Gelt is flat pieces of chocolate wrapped in gold foil and made to look like coins. I think it's underutilized. Here are some options for it:

- Crumble it onto ice cream or into breakfast cereal

- Place it inside a peanut butter or Fluffernutter sandwich

- Dip it like tortilla chips into complementary dips

THE HERETIC'S BUFFET: ADDING PORK TO CLASSIC JEWISH HOLIDAY FOODS

The purpose of these dishes is not to be offensive—just delicious.

- The Heretic's Crostini: A potato pancake topped with thin-sliced pork chop and a dab of applesauce. The pancake should be small enough so this is easily picked up and eaten by hand.

- The Heretic's Stew: Matzoh ball soup with slow-cooked roast pork.

- The Heretic's Slider: A BLT with mayo on a Passover roll. Passover rolls have a texture similar to cream puffs or profiteroles. They're all eggs and oil and they don't actually rise, because Jews don't eat leavened bread on Passover. They're pretty dry, but fill them with lettuce, tomato, mayo, and bacon and they moisten right up.

- The Heretic's Hanukkah Miracle Sandwich: The Hanukkah Miracle Sandwich described above, with bacon.

hellfires

The Heretic's Crostini

MATZOH BALL SOUP: GOOD FOR THE JEWS, AND EVERYONE ELSE

Matzoh ball soup is glorious in its simplicity and deliciousness. It's a classic comfort food and great when you have a cold, whether or not your cold is of Semitic ancestry.

At its essence, it's just a bowl of chicken broth and breadlike balls that have been engineered specifically to soak up chicken broth. Sometimes it has noodles, carrots, chicken, etc., but really, it's about the broth, the balls, and the way they come together.

That's why the best strategy when eating matzoh ball soup is to facilitate this union. Slice each ball in half to expose its interior to the broth. Spoon broth over the balls until they're beyond saturated.

If the idea of making matzoh ball soup intimidates you, just buy a mix from the store and use your own chicken broth. And here's a secret: Add a splash of club soda to the matzoh ball batter before cooking. The old Jewish ladies I know say the carbonation makes the matzoh balls fluffier.

WHY OBSERVING PASSOVER ISN'T AS BAD AS JEWS MAKE IT SOUND—AND HOW TO MAKE IT BETTER

Passover is the holiday when we Jews commemorate our exodus from Egypt. As the story goes, we were crossing the desert in a bit of a hurry and didn't have time for our bread to rise. So each year on Passover we remember that time by spending eight days not eating regular bread and instead eating unleavened bread—matzoh.

Many Jews also avoid foods like pasta and beer that use breadlike ingredients. Some go even further. The point is, Passover is generally a time when Jews get together and complain about not being able to eat certain foods. These complaints ignore two important facts:

1. You can still eat beef, chicken, turkey, fish, eggs, cheese, fruit, and almost all vegetables. That's more than vegans can eat. And this is only for EIGHT DAYS.

2. Matzoh is actually pretty good. It's like a thicker, crunchier version of a table water cracker, with a charring that adds unique flavor. It's much milder and lighter than most oily, salty snacks, so it's a wonderful dip conduit that won't overpower. For eight days a year, it's a nice change of pace.

But most Jews insist upon viewing Passover observance as some kind of cross to bear. When the holiday ends, they run frantically to the nearest pizzeria, as if throughout the rest of the year they never go eight days without eating pizza.

I want my Jewish brothers and sisters to get as much enjoyment from Passover as I do. And I believe that even the world's oldest traditions need a little shaking up every few thousand years. So now I'll highlight some of the benefits of Passover observance, and introduce some new recipes that would make a trek across the desert well worthwhile for Jews and gentiles alike.

THE SALTWATER SPRINKLE: A YEAR-ROUND HARD-BOILED EGG IMPROVEMENT

The Passover observance involves the use of salt water, often poured over a hard-boiled egg. This custom has unwittingly provided us a different, and I believe superior, way to salt hard-boiled eggs. It offers more uniform salt coverage and allows you to moisten the egg as you season it. (Fresh mozzarella is often dipped in salt water. This works on a similar principle.) Just remember that you want a gentle sprinkle, not a soaking.

PASSOVER SANGRIA

Make no mistake—Passover is a drinking holiday. During a single night's observance Jews are supposed to consume four glasses of wine as we recall the plight of our ancestors. But why not drink enough to *forget* the plight of our ancestors? After all, they don't want us to worry about them. They only want us to be happy. Can't you just hear them?

"Don't worry about us. We only want you to be happy. Go, have a good time with your friends. We'll be fine. We love building pyramids. Heavy? No, these bricks are light as feathers!"

This sangria includes some Passover staples:

- **MANISCHEWITZ WINE** A red wine with lots of sugar added.

- **CHAROSET** Represents mortar from the building of the pyramids; generally chopped apples, nuts, cinnamon, sugar, and more Manischewitz, although recipes vary widely.

- **MAROR** Bitter herbs to represent the Jews' suffering in Egypt— usually horseradish.

This recipe makes about 1½ quarts. As always, tweak it to your liking.

YOU WILL NEED

1 bottle (750 ml) Concord grape Manischewitz

1 cup Calvados (can substitute brandy)

3 cups charoset

¼ cup fresh lemon juice

¼ cup fresh lime juice

Pinch salt

4 (¼-inch-thick) slices fresh horseradish, a.k.a. maror (optional)

Seasonal Fruits (optional):

        Orange, grapefruit, or tangerine, peeled and segmented

Kiwi, peeled and quartered

Cherries, pitted and halved

INSTRUCTIONS

Combine everything in large pitcher. Cover and refrigerate at least 8 hours and up to 48 hours. Serve in wine glasses filled with ice cubes.

Notes

- Charoset that isn't too sweet and is made with Cortland or McIntosh apples is best for this recipe. If you want to get really hardcore, you can swap out the charoset for 2 chopped, peeled apples, ½ cup of toasted walnut halves, and 2 cinnamon sticks, so the charoset isn't adding extra sugar.

- If you want even more tart, substitute grapefruit juice for lemon and/or lime juice. Orange juice is also a nice option.

- Peeled horseradish needs to sit exposed to the air for at least 10 minutes in order for it to develop its characteristic spiciness.

MANISCHEWITZ SORBET

This one seems so obvious to me, I'm embarrassed on behalf of Jews worldwide that it's not already everywhere. If you like the idea of red wine sorbet, you'll probably like this.

Mix as much as you want to make in proportions of three-quarters Manischewitz and one-quarter grape juice. Pour into a baking pan and freeze overnight. Scrape into a serving container with a fork. The result is the perfect Passover palate cleanser. Just don't forget that it still contains alcohol.

EATING AT WORK

I t's time to study our culture by moving from the exceptional to the mundane. Because while holidays may get starred on the calendar, everyday eating has a larger cumulative effect on our pursuit of Perfect Deliciousness.

ON OPTIMAL CUBE FARM CUISINE
(SPOILER ALERT: IT'S NOT SANDWICHES)

T he cube farm is the great petri dish of office culture, so it's important to understand how to eat there. If you work and eat on the go, sandwiches are likely the best lunch for you. But if you sit and eat at a desk, the sandwich is an overrated vessel.

To be clear, it's a good option, just not as good as people think for this purpose. The sandwich's quintessential characteristic—that it's eaten with the hands—means you're likely to get your mouse, keyboard, and phone dirty as you partake in some lunchtime web surfing or texting. Other alternatives also have issues. Foods requiring a knife are unwieldy, and options with a high splatter factor, like pasta, are too dangerous.

The best food for a desk-based lunch comes in small pieces and can be eaten with a single utensil out of the container in which it was transported. That's why dishes with chopped meats and/or vegetables mixed with a base like rice or Israeli couscous make perfect cubicle lunches.

Put it in a plastic container, allow it to come to room temperature throughout the morning, heat it briefly in the microwave if you want it hot, then eat it straight from the con-

> TIP There are limits to aromatic liberty in a cube farm. (You might want to hold off on eating microwave popcorn with canned tuna on top.) Remember that just because it smells good to you doesn't mean it smells good to someone else. However, if it smells good to you and not to someone else, it may mean the other person is hard of smelling.

tainer with a spoon, or better yet, a spork. (Spoons are vastly superior to forks for these types of dishes. Why struggle to keep rice or peas on a fork when a better option is right in front of you?)

WORKPLACE KITCHENS: WHAT GOOD EMPLOYERS PROVIDE

MUSTS

- A decent coffee machine with real milk (no non-dairy creamer), sugar, and artificial sweetener
- Refrigerator
- Microwave
- Tap water filtration system (as opposed to bottled water, which is a wasteful blight on the Eatscape)
- Coffee mugs and water cups
- Dishwashing sink with sponge and soap

NICE ADDITIONS

- Toaster oven
- Espresso machine
- Freezer
- Plates, bowls, and utensils
- Tea
- Salt and pepper

GOING THE EXTRA MILE

- Olive oil
- Honey
- Fried onions
- Omelet chef

VENDING MACHINE DECISION TREE

Like the water cooler, the vending machine is an office gathering place with great cultural significance. An unspoken rule dictates that around three or four in the afternoon, workers rise from their desks and saunter toward the snacks, sometimes not even realizing until they get there that they've gotten up at all.

When you find yourself in that situation, it's important that you make a good choice. This flow chart will help:

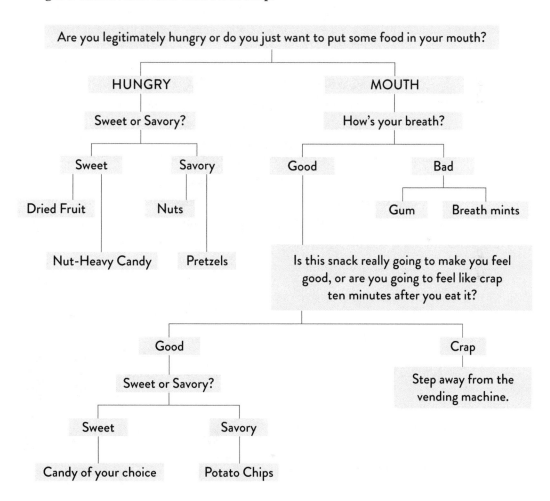

ADVANCED VENDING MACHINE CUISINE

Your office vending machine doesn't have to be the place you go when you can't bear to sit at your desk anymore. It could be the place you go when you want to soar, to join together with your coworkers and reach for the zenith of Perfect Deliciousness. With a bit of imagination you'll come to see this rickety behemoth less as a culinary dead end and more as a palette with which to thrill your palate.

There's a lot of variation among vending machine options, so these particular recipes may be difficult to reproduce precisely. But I'm confident that the concepts and techniques here will prove both instructive and inspirational.

CUBE FARM TIRAMISU

This dish is decadent enough to make even fluorescent lighting romantic—and it's especially naughty if you're using free office coffee in it.

Separate Oreos so you have creamed and uncreamed halves. Set aside uncreamed halves. In a flat-bottomed bowl or baking dish, make a single layer of creamed Oreo halves, cream side up, dunking each in coffee for 6 seconds before placing it down. Next, form a layer of Milano cookies, also each dunked in coffee for 6 seconds. Repeat with a second layer of creamed Oreo halves and Milanos.

Finished Dish

Crumble uncreamed Oreo halves and set a quarter of them aside. Mix the rest with coffee to make a paste, with some cookie texture remaining. Spread paste on top of tiramisu. Cover the tiramisu with plastic and put it in the fridge all day or overnight. When you're ready to eat, remove from fridge for 10 to 15 minutes to take the chill off. Top with whipped cream and remaining Oreo crumbs.

Many Eaters helped test recipes and techniques for this book, but one family of testers went above and beyond in their work on this recipe. Special thanks to Laura and David Schultz of Wilmington, North Carolina, and their kids Alex and Annabel—both presidents of the Sporkful Junior Eaters Society. They helped take this concept to the next level.

CHUCKLES GAZPACHO

Chuckles are a gummy, sugary, fruit-colored treat, but if you can't find them near you, anything along those lines will work. Pour Chuckles and lemon-lime soda into a bowl. Let them marinate for a few minutes, then enjoy. When eating, you may want to hold a Chuckle between your teeth and work the broth through it. To decrease sweetness, use club soda.

SHEET CAKE SLICING TECHNIQUES BY OCCASION

Sheet cakes are an important part of ritual observances in office culture. These cakes have three types of slices: corner, edge, and middle. The difference between the three is the quantity of side frosting, and thus the level of decadence. The type of slice given to the guest of honor depends upon the occasion.

- **BIRTHDAY** Allow the birthday boy/girl to choose.

- **VOLUNTARY EXIT** A treasured coworker who's moving on deserves a corner piece. An insufferable rival gets nothing.

- **RETIREMENT** Honorees who seem near death should be given a large corner. Those with a few more years should be given a small middle.

- **FIRING** These people could be loose cannons. It's best to toe the middle ground—stay in their good graces without looking disloyal to the boss. Slip an edge slice into their packed box of personal belongings.

- **PREGNANCY** This is a tough one. If you ask the pregnant woman to decide what she wants, you could be there all day. If you decide for her, you may choose incorrectly. Want to indulge her with a corner piece? Better hope her doctor didn't just tell her to watch her weight gain. Think it's best to go with a middle? You just called her fat. Confident that an edge is a safe happy medium? You're so insensitive!* Your best bet is to get a round cake and let the guest of honor cut her own slice.

CONSUMPTION IN THE CAR

Few activities are more American than driving and eating. Doing them well together is integral to our nation's folklore and culture, as well as to eating more better.

DRIVING + EATING = AMERICA

To be clear, eating behind the wheel is suboptimal. It's a degraded dining experience and an unsafe way to drive. But sometimes it has to be done. If you're going to endanger yourself, your passengers, and everyone else on the road, please, at least make sure that your (possibly) last meal is a good one.

- DO prepare your meal station before you start driving. Apply condiments and prep your fry-dipping area. Put foods you wish to access on the passenger seat or in the cup holders.

- DON'T put it on the floor or your lap. Ducking down to reach the floor takes your eyes too far from the road, and if it's on your lap and you have to move your legs suddenly as you drive, that burger is a goner.

---

* Please note that Mrs. Sporkful read this paragraph while pregnant and approved its publication.

FAST FOOD FEASTING TIPS

- Request a small customization to a standard preparation—like no onions on a burger—to force them to make a fresh one instead of taking one from under the heat lamp.

- To get your food faster, it's sometimes better to order at the drive-thru, then eat inside. The employee at the drive-thru is often the most skilled and experienced on the shift.

- If you're using the drive-thru and planning on eating in the car, send someone inside to gather condiments while you wait for your order. This speeds the process and ensures adequate condimentation.

- DO eat the fries right away. They have an extremely short life expectancy, especially once they've been enclosed in a bag and steam turns into condensation. You should really never order take-out fries unless you plan to eat them on the way home.

- DON'T squirt ketchup directly onto the fries. It's too messy, and anyway, fries should be dipped on a per-bite basis at all times, to maintain crisp and regulate ratios.

- DON'T feel obligated to dip the fries in any condiment at all.

- DO place hard plastic condiment containers in a cup holder.

- DON'T eat foods that require utensils. Ever.

- DO use the bag for garbage, not as a placemat or napkin.

A burger or sandwich in a wrapper is slightly better than one in a carton, because the wrapper can be spread out to protect the passenger seat and the edges folded up to catch wayward crumbs. However, if your condiments are in soft pouches instead of hard plastic containers, the carton does provide a more reliable condiment container and dipping basin.

The Eater's greatest predicament occurs when you have soft condiment pouches and a paper wrapper, and you want to dip your fries in something. What do you do with the condiment? You can form a small mound of it on the wrapper on the passenger seat, but that's perilous from the start. As an alternative, you may get some extra napkins and place the burger on those, then fashion the wrapper into a dipping basin and cup holder protector (figure 6.1).

Fig. 6.1

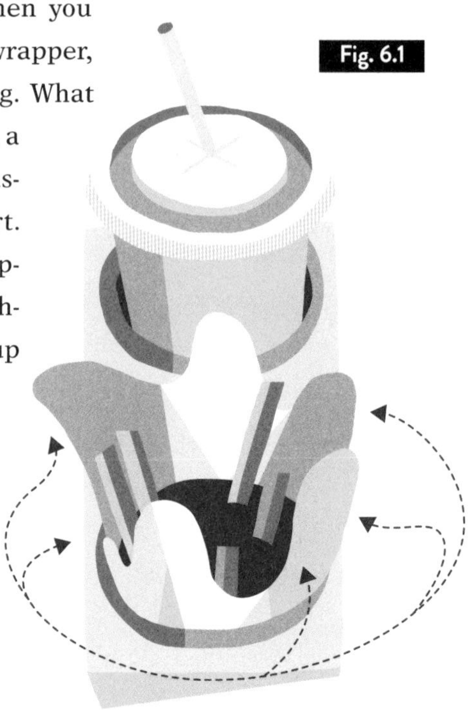

Makeshift Vehicular French Fry Dipping Basin

Leave part of the wrapper sticking out of the cup holder so you have something to grab when you're ready to throw it out.

DRIVER AND PASSENGER AREAS OF RESPONSIBILITY

Drivers must transport passengers safely, and passengers must help drivers eat safely. It would be as wrong to taunt a driver with deliciousness just out of reach as it would be to taunt passengers by wrapping your car around a tree.

While drivers focus on their task, passengers can reciprocate by handing a sandwich to the driver, pointing out choice bites and holding it between chews. Dip a fry in the driver's condiment of choice and pass it over, uncondimented end first, so it can be grasped easily and cleanly. Offer beverages as needed. Beautiful customs like these help to perpetuate the American experience.

HOMEWORK

To improve the Eatscape, we must continue to study not only the types of foods our culture holds aloft but also how those foods attain such status. Once we understand that, we'll be able to build upon tradition to create new and better customs. Which brings me to your assignment . . .

As host of *The Sporkful*, I get lots of pitches from PR people, many of which are pegged to celebratory days I never knew existed. One especially decadent week in July apparently includes National Corn Fritter Day, National Caviar Day, and National Daiquiri Day. (They could save us some time by just declaring it National Caviar Corn Fritter Daiquiri Week.)

Most of these "holidays" were invented by the same PR people sending the pitches, to give them an excuse to get members of the media to talk about the foods and drinks they're promoting. This is an interesting commentary on our culture that leads to an obvious conclusion: We Eaters need our own day!

Your assignment, then, is to help me figure out how we should observe Eaters Day, a worldwide celebration of our never-ending quest for deliciousness. I'd also love to get suggestions on what day it should be, as it seems most of them are taken. Send me your thoughts at dan@sporkful.com. And happy Eaters Day!

Submit your homework to me at dan@sporkful.com.

7

MATHEMATICS
Calculating Your Way
to Deliciousness

- Pie aren't squared—but it should be

- The geometry of sandwiches, potato chips, and cocktail garnishes

- Ideal ratios between bread and filling, bagel and spread, hot dog and bun

- Fibonacci microwave sequences

- Mozzarella's golden ratio

- Exponential eating

- Menu factorials

- The Cobb salad equation

- Algebraic recipes

Math has long been a target of scorn among students.

"When will we ever use this?" they cry.

"We have machines to do this for us now," they carp.

"You forgot to carry the one," they tell me.

Math can be intimidating. The numbers swirl around you, mocking your feeble attempts to corral them, and a feeling of helplessness ensues. I understand. I spent my formative years avoiding the subject.

To circumvent my college's math requirement I took two nearly identical introductory computer science courses at the same time, because they both counted toward the requirement. I asked a professor, "Can I take Comp Sci Five and Six in the same semester and be done with my math requirement forever?" And he said, "Well, I guess technically we can't stop you."

That was all I needed to hear. (I still don't know what a HyperCard is, or how to use one.)

Despite how hard I worked to avoid math in college, I'm thankful for what I learned of it in high school, because I see now that it applies to so many areas of life.

In other words, perseverance pays off. With focus and discipline in your math studies, numbers become your servants, and life's seemingly random events align themselves neatly along X and Y axes. This enlightenment leads to discoveries like the one made by performance artist and musician Gary Wilson, who understood that "6.4 = Make Out."

In the realm of food and eating, math can be just as revelatory. In this chapter we'll apply the lessons of geometry to consider a food's ideal shape, calibrate sandwich ratios, and use exponents to multiply deliciousness many times over.

As we progress through increasingly complex concepts, you'll see that my attitude toward math has changed so dramatically that I've not only learned to appreciate its study, I've also stopped resenting people who are good at it. At the culmination of this chapter, I'll reach out to those mathematically inclined Eaters among us, by putting principles of eating into a language they too can understand.

And you thought this subject was useless!

GEOMETRY

Geometry is the study of shapes, a subject that affects every meal you'll ever eat. The shape of food is something so fundamental to your eating experience that you may never have thought to question it. But we Eaters do not take geometry for granted.

Why are foods shaped as they are? Would they be more delicious if they were shaped differently? And when a food comes in different shapes, which variety is best, and why? Let's find out.

PIE AREN'T SQUARED—BUT IT SHOULD BE

Eaters long ago observed that no matter how big a pie is, the ratio of its circumference to its diameter remains the same. We use this special number to calculate the area of a pie, which leads us to utter the magic formula "Pie are squared."

Unfortunately, this statement has yet to influence pie baking across the greater Eatscape. Together, we can change that.

Pie is generally made up of two components—filling and crust. But while the filling is consistent throughout, the crust's properties change by region (see figure 7.1).

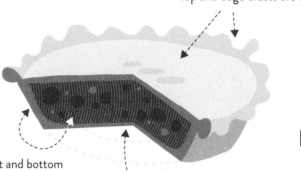

Top and edge crusts are often dry.

Fig. 7.1

Vertical side crust and bottom outer crust are the happy medium—moist and sturdy.

Bottom middle crust is often too thin to stand up to fillings.

In the best pies,* each region of crust makes a positive contribution, adding its own texture and flavor, as well as absorbing some of the filling. But more often, top and edge crusts are dry, and bottom inner crust is so thin that it dissolves upon contact with the filling's moist advances. Bottom outer crust and vertical side crust are superior, because of their positioning and thickness. They're ideally suited to absorb flavor from the fillings without turning to mush.

In a typical slice of pie, you'll find the ideal bites precisely where these two crustular regions come together with each other and the filling, in an arrangement that French mathematicians have dubbed a Ménage à 3.14159 (figure 7.2). There can be no doubt that pie would be better if the ratio of optimal crusts to suboptimal crusts were increased. And that's exactly what happens when pie are squared.

TIP Of course, if you make a large square pie and cut it up like a tic-tac-toe board, some pieces will have more or less crust, and the center piece will have no ideal bites at all. Make two small square pies instead, so each can be cut into four identical square servings. This promotes harmony and equality among Eaters.

---

* "The best pies" means those made with a lard crust.

Ménage à 3.14159 and the Superiority of Squared Pie

Wherever vertical side crust meets bottom outer crust and the two are in congress with pie fillings, you have an ideal bite, a Ménage à 3.14159. A square pie offers more of these bites, because it has more perimeter than a circular pie with the same area.

Square pie will improve countless lives, but it's not without opponents. Companies that produce round pie tins, pie plates, pie servers, pie crusts, and pie charts don't like it one bit. Nonetheless these findings are sound, and I expect the unmistakable will of the people to prevail.

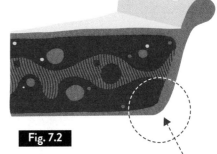

Fig. 7.2

PIE CRUST DIPPERS

When confronted with dry edge crust, remove it and insert into the middle of the pie (or the pointy tip of a slice), which tends to be an overmoistened and undercrusted area. Let the crust marinate in the filling before eating both together. Or simply break off edge crust and dip it into the central pie filling. This will have a deleterious effect on the pie's appearance but a delicious effect on its taste.

PORTION MORE PRECISELY BY SLICING PIE IN DIGITS

There's often confusion when slicing pie for others. Someone says, "I only want a little," but you don't know exactly what that means. That's why we should all state our slice size preferences in digits. You've heard of three fingers of whiskey? You can order three digits of pie. (The slicer then holds the requested number of fingers over the perimeter of the pie to measure the width of the slice.)

High-level mathematicians have already adopted this method and have calculated ten trillion digits of pie, though they still haven't eaten most of it.

EATING RIGHT TRIANGLES RIGHT

When you cut a typical white bread sandwich in half, you have two options: Slice vertically, resulting in two rectangular halves, or slice diagonally, to get two right triangles. Which offers a better eating experience, and why?

The right triangle is superior. It's more aesthetically pleasing, better-weighted in the hand, and exposes more of the sandwich interior to the eye, allowing for more informed bite selection. Plus, it offers the Eater two acute angles into which to bite.

**The Right Triangle Sandwich Half:
Advantages of the Diagonal Cut**

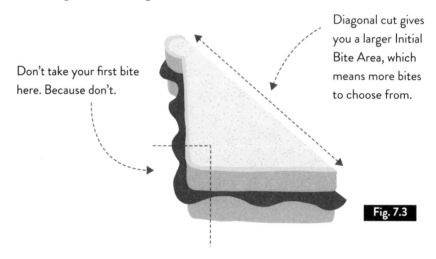

Don't take your first bite here. Because don't.

Diagonal cut gives you a larger Initial Bite Area, which means more bites to choose from.

Fig. 7.3

When *Radiolab* cohost Robert Krulwich came on *The Sporkful*, he agreed that the right triangle is superior, but we didn't see eye to eye on where the first bite from such a half should be taken—the hypotenuse or one of the acute angles. Robert advocates the hypotenuse at all times, in spite of the fact that it may besmirch one's cheeks (see figure 7.4). I believe it's sandwich dependent.

Taking First Bite from Hypotenuse Increases Risk of Cheek Mess

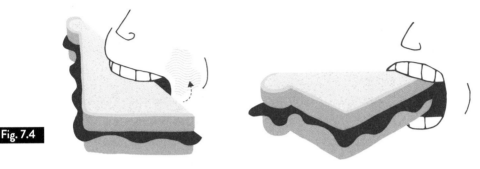

Fig. 7.4

Cheek-to-sandwich contact = messy   No cheek contact = less mess

Krulwich makes his case:

> *It's like a kiss. If you want to kiss somebody, would you pucker, pucker, pucker, and stretch your lips so you look as much like a duck as possible, and then take a little peck? No. You would bury yourself in the lips of the other. A sandwich is a form of comfort, and addressing with ardor the meal. You hit that sandwich big-time. You can't choose the acute angle. That's the sandwich's way of spiking, of saying, "No, not here!"*

Other Eaters would argue, of course, that a sandwich's point is an instructive arrow gesturing mouthward rather than a warning. In any event, I maintain that the position of your first bite should depend on the sandwich. Here are the situations that call for starting with an acute angle:

- If the sandwich has a large volume of moist or sticky ingredients or spreads, like peanut butter and jelly, that's a sign that approaching from an acute angle may keep the cheeks clean.

- If it looks like a bite in the hypotenuse might force fillings out the back, that's a cosign in favor of starting with the acute angle.

- If you ignore both the sign and cosign and end up with your face covered and/or your sandwich in your lap, thus distracting you from the conversation at hand, the new topic is a tangent. When that happens you should exclaim, "Soh cah toa!"

OPTIMAL POTATO CHIP GEOMETRY

The line between great potato chips and poor ones is as thin as a chip itself. All potato chips are crunchy and salty, but the lesser among them are nothing more. Great chips offer authentic potato flavor and place a high priority on crunch maximization. To these fine specimens, salt's just a seasoning while crunch is a creed.

It should not surprise you, then, to hear that I consider kettle-cooked potato chips to be superior. But some are superior in their superiority. When faced with a bag or bowl of kettle-cooked chips, target these shapes:

DOUBLE FOLD Most crunch per square inch of spud, period. Eat a few of these in a row and you'll have potato chips ringing in your ears.

SINGLE FOLD Less crunch than the double, but more surface area to bite into. And while I think good potato chips need no dip, the single fold offers the best combination of crunch and dippability.

STARBURST This chip captures the imagination with its complex beauty and extraordinary crunch potential, but its consumption can be fraught. Vertical insertion may lead to cutting on the roof of one's mouth, a condition known as Cap'n Crunch's Complaint. (To avoid this fate insert it sideways into your mouth.)

PIPELINE It offers a unique crunch, perhaps best described as a shattering. You can fill the cavity with dip, and when empty, it may be used as a telescope with a magnification rate of 1:1.

EYE OF THE STORM These are the diminutive flat chips that end up cooked medium-well because of their small stature. They attain a darker brown color, charred potato flavor, and light crisp. They're the counterpoint to the above shapes, and thus the ideal change of potato pace.

PROS AND CONS OF CITRUS GARNISH SHAPES

Drinks often come with a citrus garnish. The degree to which that fruit flavors your beverage, and vice versa, can vary widely depending on the shape of the garnish and how you use it. Your choices won't just influence the drinking experience, they'll also affect the garnish-eating experience. This guide will help you decide which citrus shape is right for you.

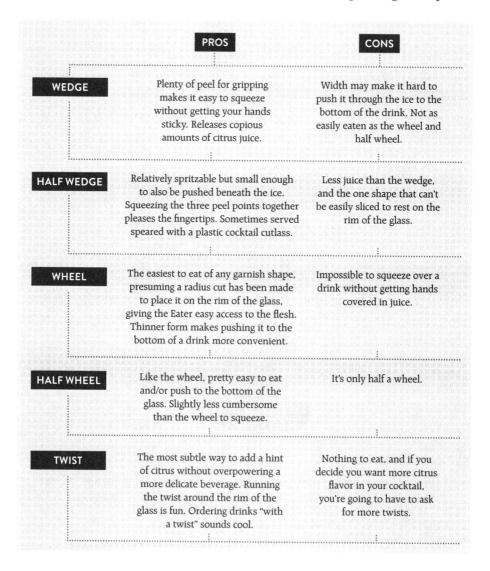

	PROS	CONS
WEDGE	Plenty of peel for gripping makes it easy to squeeze without getting your hands sticky. Releases copious amounts of citrus juice.	Width may make it hard to push it through the ice to the bottom of the drink. Not as easily eaten as the wheel and half wheel.
HALF WEDGE	Relatively spritzable but small enough to also be pushed beneath the ice. Squeezing the three peel points together pleases the fingertips. Sometimes served speared with a plastic cocktail cutlass.	Less juice than the wedge, and the one shape that can't be easily sliced to rest on the rim of the glass.
WHEEL	The easiest to eat of any garnish shape, presuming a radius cut has been made to place it on the rim of the glass, giving the Eater easy access to the flesh. Thinner form makes pushing it to the bottom of a drink more convenient.	Impossible to squeeze over a drink without getting hands covered in juice.
HALF WHEEL	Like the wheel, pretty easy to eat and/or push to the bottom of the glass. Slightly less cumbersome than the wheel to squeeze.	It's only half a wheel.
TWIST	The most subtle way to add a hint of citrus without overpowering a more delicate beverage. Running the twist around the rim of the glass is fun. Ordering drinks "with a twist" sounds cool.	Nothing to eat, and if you decide you want more citrus flavor in your cocktail, you're going to have to ask for more twists.

BOOSTING GARNISH FLAVOR

Lemons, limes, and the like should always be served on the rim of a glass, un-squeezed, so that the Eater has omnipotence in determining the level of citrus flavor in a drink. I'm sure you're familiar with the basic ways of doing that, but if you want to get the most possible citrus flavor into your beverage, you're going to need this more advanced multipart maneuver.

Step 1: Squeeze as much juice from the citrus into the drink as possible, then drop the fruit in. **Step 2:** As you stir the drink gently, use your straw to plunge the fruit down to the bottom of the glass. Pin it beneath the ice so it stays there. **Step 3:** Position your straw at the bottom of the glass, next to the garnish, and begin drinking.

RACHEL MADDOW: "THE GARNISH IS NOT A SNACK." ME: "YES IT IS."

When appearing on *The Sporkful*, MSNBC's Rachel Maddow said, "You're never supposed to eat the garnish. The garnish is not a snack. The garnish is there to modify the liquor that is in the glass. It is not there to provide you a tasty treat to go along with your drink. The drink is the treat."

With all due respect to Ms. Maddow, a thoughtful and passionate Eater, I must disagree. The primary purpose of a garnish is to impart flavor to a drink, but the benefits that a drink imparts to a garnish must not be trivialized. A boozy fruit can be as delightful and timely as fruity booze, and I see no reason to deprive ourselves of either.

The real question is, when is the garnish best eaten? I like mine when the drink is about two-thirds or three-quarters finished. Neither the drink nor the garnish is likely to exchange much more flavor, and the fruit offers a nice change of pace for the palate before you return to polish off your beverage.

RATIOS AND FRACTIONS

Ratios affect deliciousness across disciplines. The ratio of the components of a food to each other is often more important than the components themselves. Just ask anyone who's ever failed to melt enough cheese on their caviar.

I'll go over some essential ratio principles now, and we'll come back to the discussion at various points throughout this book. It's my hope that a firm understanding of this subject will give you the foundation to make wise choices in situations we may not have discussed. This is what we call "Education for Life."

ON THE RATIO OF SANDWICH BREAD TO SANDWICH FILLING

When making a sandwich, ask yourself:

1. How high will the fillings be once piled on? If the answer is "Very high," keep the bread sturdy but thin. You don't want your structure elevated to inedible heights.

2. How strong in flavor are the fillings? You'll have no trouble tasting the bread, because it's the first and last part of each bite. But Eaters who pair thick bread with a mild interior may have difficulty tasting some components.

3. How moist are the fillings? Choosing thicker bread is one way to ward off sogginess, though not as effective as using a crusty roll, as detailed on page 78 of the "Engineering" chapter, along with more applied sandwich mathematics.

TIP Whoever coined the term "the best thing since sliced bread" either didn't put a high priority on ratios or didn't get out much. When you purchase bread unsliced and cut it as you need it, you can alter slice thickness depending on the ratio a given sandwich requires. For more on this issue see "Sliced Bread: The Worst Thing Since Itself," on page 11 in "Physical Sciences."

A BAGEL CRISIS RESOLVED

A confluence of trends has created controversy in the bagel community. Bagels are getting bigger, losing exterior crustiness, and puffing up to the point that their trademark holes are closing, blurring the line that separates them from rolls. Will purebred bagels be mongrelized to the point of extinction?

I believe reports of the traditional bagel's demise are greatly exaggerated, but this issue does have grave implications for bagel sandwichization, because of its impact on sandwich bread-to-filling ratios. Coupled with the anticarb craze of recent years, this bagelflation trend has led to two developments:

1. The emergence of the "flagel," a flat, bagelesque product that can be quite enjoyable, but could never adequately substitute for a true bagel; and

2. A sharp increase in bagel scooping, a vulgar method of ratio alteration and/or portion control, which has the unfortunate side effect of creating a "butter gutter," as NPR's Ian Chillag called it on *The Sporkful.*

The central problem with both flagels and scooping is that they remove almost all of the bready interior, rendering the bagel something different—something less. So how best to combat bagelflation and bring sandwich ratios in line, without resorting to bagel barbarism? Bagel Trifurcation.

TRIFURCATED BAGEL

When sandwichizing a bagel, slice it horizontally into thirds, instead of halves. Remove the middle third and set it aside. You've now reduced the height of the bagel and improved its ratio to the filling, while preserving a good portion of luscious interior.

Lest you think this recipe would waste any bagel, note that the middle third you just removed offers great potential. Toast it and use it as a homemade bagel chip with the spread of your choice, or slice it in half and make a separate half sandwich. If you want to eat well and learn about fractions at the same time, make Dodeca-hedron Dippers (see recipe below).

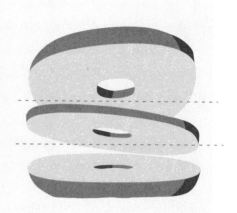

So come down from that ledge, bagel sandwichizers, and retire your scooping fingers. There is a better way. And to those of you who insist upon continuing to scoop: The only thing worse than doing that even after having read about Bagel Trifurcation is wasting the morsels of bagel dough you insist upon scooping. Toast them on a baking sheet in the oven and use them in a bread salad, like *panzanella* or *dakos*.

DODECAHEDRON DIPPERS

If you or your child is struggling with fractions and/or hunger, this recipe will help. Prepare the dip or dips of your choice. Trifurcate 2 bagels as previously described, resulting in 6 bagel thirds. Lightly brush both sides of each bagel slice with olive oil and sprinkle with salt and pepper to taste. Broil the 6 thirds until golden brown, about 1 minute on each side. While they're still hot, rub the thirds with the sliced side of a head of garlic.

Cut each bagel third in half, to make sixths. Cut the sixths in half again to make twelfths. Because you started with 2 bagels, you should now have 24 twelfths, which equals 2. That is math.

BAGEL HALVES AND HALVE-NOTS
(THE JACK SPRAT COROLLARY)

Bagels are often served already sliced, and Eaters who only want half a bagel then take either a top or a bottom and leave the rest. This is problematic for two reasons.

First, you're depriving yourself of the textural variety offered by the top and bottom bagel halves. If you want a full-fledged bagel experience, slice the entire halved bagel in half vertically, then take a top quarter and bottom quarter. This way you have a true half bagel, easily sandwichized or eaten open-faced.

If you happen to prefer the top or bottom, that's fine in a vacuum, but understand that by only taking one or the other, you're leaving an incomplete half for a fellow Eater. In that case, you should make sure someone else there prefers the opposite half. This is an example of the Jack Sprat Corollary, named for a man who could eat no fat and whose wife could eat no lean. (Together they finished 100 percent of their food.)

IMPROVING BUTTER PROPORTIONALITY
THROUGH MUFFIN TRIFURCATION

A buttered, griddled muffin is a wonderful thing. The muffin's cakey consistency is ideal for butter absorption, and adding more crisp through griddling means the muffin's top isn't the only place you'll find great texture. So have more of that, with Muffin Trifurcation!

Slicing a muffin in thirds vertically exposes more muffin interior to butter and griddle, increasing all the blessings of this delicious food. And by slicing vertically, you ensure that each piece still has some muffin top.

"BUN SIZE" IS NO SIZE FOR EATERS

It may be mathematically dubious, but I maintain that at least three-fourths of the hot dog buns sold in America are below average. They're dry, flavorless, and flimsy, and they get in the way of hot dog enjoyment. That's why I object to the emphasis placed on finding franks and buns that are equal in length (a 1:1 ratio). (Some hot dogs are even marketed to the unknowing as "bun size.")

Allowing for a few bites of bunless frank is a sign of wisdom, not misman-agement. The dog is the star, its flavor and texture paramount, especially if yours is blessed with natural casing. If you've got a bum bun, the only positive contribution it's making is as a grip.

When eating a quality hot dog on an ordinary bun, trim the bun to achieve a 2:1 dog-to-bun ratio, or just buy half as many buns as dogs and cut the buns in half. Rather than spread the condiment on the bunless regions of frank, which sends your Stainability Quotient through the roof, dip the dog on a per-bite basis. (There's more on dipping on page 229 in "Psychology.")

FIBONACCI AND THE GOLDEN RATIO

The Italian mathematician Leonardo of Pisa, better known as Fibonacci, made contributions to his field that remain relevant today. But even he did not realize how his work would influence the Eatscape nearly eight hundred years after his death.

Fibonacci did a lot of work with the golden ratio, a relationship between two numbers whereby the ratio of their sum to the larger number is equal to the ratio of the larger number to the smaller one. It's best illustrated with this drawing, in which the ratio of the size of box A to the size of box B is the same as the ratio of the two boxes combined to box A. (That ratio is about 1.6:1.)

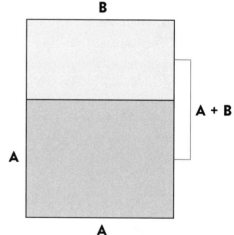

The golden ratio comes in handy when you're deciding how much fresh mozzarella to use in a sandwich. Most cheeses are supporting sandwich actors, but because of fresh mozzarella's mild flavor and toothsinkable texture, and because when made well it is one of the most golden foods of all, it should more often be treated as the star.

When you make a sandwich with fresh mozzarella, make sure it's done in the golden ratio, where the mozzarella is portioned as the larger quantity and its accompaniment is the smaller quantity. This will make sure the featured attraction—the cheese—is not overshadowed.

THE FIBONACCI SEQUENCE APPLIED TO MICROWAVE USE

Fibonacci's work with the golden ratio was far from his only contribution to mathematics. He is also the man behind the Fibonacci sequence, a list of numbers starting with zero or one such that each successive number is the sum of the previous two (1, 1, 2, 3, 5, 8, 13, 21 . . .). Of course, he didn't live to see the microwave, so we can only assume that he would have been happy to have his sequence employed as a guide for its use.

I recommend using the microwave sparingly, in eight-, thirteen-, or twenty-one-second bursts. These durations are especially handy for taking the chill off foods from the fridge when you don't have time to wait for it to happen naturally.

Here are several examples of uses for a Fibonacci microwave burst, along with the recommended length of the burst in each case:

8 SECONDS to soften butter for spreading or take the chill off fruits or salad greens from the fridge, thus restoring their flavor without cooking them.

13 SECONDS to soften a pint of ice cream that's too hard to scoop and eat, when you can't wait for it to defrost because you want to eat it now.

21 SECONDS to get chilled fresh mozzarella just warm enough to border on melty but still firm enough to be picked up by hand, or to heat leftover steak that's already been brought to room temperature, to get it warm through without cooking it more.

Microwaves vary widely, so adjust these times to work for you. The point is that your microwave can be used in short intervals to make subtle alterations to a food's temperature. Just as Fibonacci intended.

EXPONENTS

Eaters use exponents to multiply foods by themselves, in order to make meals exponentially more delicious. Exponential eating can be as simple as dipping an orange wedge into orange juice, as my daughter is fond of doing. But it can also be more complex.

Exponents are represented with a small superscript number equal to the number of times you're multiplying the food by itself. For example:

Pizza x Pizza is written as Pizza^2 (Pizza Squared).

While exponential eating offers great promise, it should be noted that just because a food is squared exponentially does not mean it must be squared geometrically.

PIZZA^2

This recipe introduces a new and delicious pizza topping: pizza. Cut a slice of pizza into small squares about ½ inch on each side. Set crust aside for later munching. Sprinkle squares on another slice of pizza and enjoy.

PIZZA SCISSORS AND FOOD FLATTENERS

If you've been to a two-year-old's birthday party recently, you know that many parents have given up using knives to cut pizza into small pieces for their tykes, opting instead for scissors. The technique is quite effective, but while any old scissors will do the trick, some unscrupulous manufacturers have sought to prey upon the chopping challenged by selling "pizza scissors."

Good kitchen shears have their place, but anyone who spends $30 on high-end pizza scissors probably deserves a kitchen full of pointless appliances. And might I add that this book makes for an excellent food flattener. Tired of foods that are too high to fit in your mouth? Buy this book!

TORTILLAS² (OR, THE TREAT OF GUADALUPE HIDALGO)

Corn and flour tortillas are rivals, but this dish capitalizes upon their complementary strengths. Together, they form Tortillas², also known as the Treat of Guadalupe Hidalgo.

Place a flour tortilla on a plate, sprinkle with shredded cheese, and microwave until cheese is melted. Place corn tortilla chips across top half of flour tortilla so each chip points to the center of the tortilla:

Fold flour tortilla in half, bringing bottom edge up to the top. Place a double layer of chips on one side of the folded tortilla, then fold in half again:

The second batch of chips is not fastened with melted cheese because you don't want to put the structure back in the microwave to melt more cheese once you've added the first layer of chips—they'll turn soggy. As long as the flour tortilla is warm and soft, it will envelop the chips and help maintain structural integrity. As you fold the tortilla over, don't be alarmed if the chips crack a bit as you press it all together.

Consume immediately, plain or with the dip(s) of your choice. Once you've mastered the technique, you can swap in different chips or other sources of crunch as you like.

BONUS RECIPE: TORTILLAS[3]

In Spain, a tortilla is not a flat doughy thing but a dish like a frittata, or thick omelet, with potatoes inside. (You can also make it with potato chips instead of fresh potatoes, à la chef Ferran Adrià, a master of molecular gastronomy who's famous for his work with potato chips, among other foods.) To make Tortillas[3], simply insert thin, triangular slices of Spanish tortilla along with the tortilla chips in Tortillas[2]. You now have Tortillas[3].

OMEGA-3³

Place a pan-seared salmon fillet on a plate, top with a single layer of thin-sliced smoked salmon, and top that with a slice of crispy pan-fried salmon skin. Sprinkle with scallions. (Add salmon roe to make Omega-3⁴.)

You could make this dish look prettier by putting the smoked salmon on the bottom, but then it would end up on the tip of your fork and be the first thing to hit your tongue. You don't want the saltiest, strongest-flavored component in that position, or it will overshadow other flavors. The salmon skin is better on top because if you bury it under moist and/or warm ingredients, it will lose its crisp. By stacking the ingredients as described you'll begin the tasting experience with the subtler fillet, then get a nice hit of saltier smoked salmon, then finish it with the texture and fat of the skin. Glorious.

TWICE-BAKED POTATO² SANDWICH

Turn a twice-baked potato over and place it on top of another, then eat them as a sandwich. For texture and flavor add something crunchy in the middle, like bacon, scallions, or fried onions.

ALGEBRA AND ADVANCED MATH

For most of this chapter, I've tried to reach Eaters by putting mathematical principles into terms we can all understand and use. But as we get to the most complex concepts I'm capable of covering, I'd like to turn the tables and put eating principles into terms that the mathematically inclined can use. In so doing, I hope to build a bridge to those who may be great at math but not as good at eating, so we may come together and grow the ranks of the Eatscape.

EXPANDING MENU OPTIONS THROUGH FACTORIALS

Factorials can be used to determine how many different ways you can arrange a certain number of ingredients. They're especially useful for downtrodden Eaters who feel the four items remaining in their pantries doom them to disappointing meals.

Factorials are represented with a !, so when you have four ingredients and want to know how many different ways you might arrange them, you write, "4!" But there's no need to shout, unless you want to calculate a factorial loudly, in which case you write, "4!!" Of course, if you're calculating a factorial of Mexican foods loudly, you write, "¡4!"

To find a number's factorial, multiply all the numbers from one to the number in question by each other. The math looks like this:

$$4! = 4 \times 3 \times 2 \times 1 = 24$$

So if a paltry pantry has you down, realize that with only four ingredients, there are twenty-four possible dishes you could prepare! (Technically there are only twenty-four if changing the order in which you combine ingredients changes the dish. Sometimes it does and sometimes it doesn't. But the basic point remains the same: Your kitchen is a lab, and this is your time to experiment. Four available ingredients is not a problem, it's an opportunity.)

ORDER OF OPERATIONS

In solving algebraic equations, mathematicians follow a proscribed order of operations. (Remember "Please Excuse My Dear Aunt Sally?" Neither do I.) Cooking also has an order of operations, better known as a "recipe." Here I've converted the instructions for several timeless dishes into the language of math.

ALGEBRAIC PIZZA

Remember that in algebra, parentheses are used to group parts of an equation, and when a figure is placed just outside the parentheses, each figure inside the parentheses is multiplied by it. For instance:

$$5(a + b) = 5a + 5b$$

Algebraic Pizza is represented as follows:

$$\text{Heat}[(\text{flour} + \text{water} + \text{yeast} + \text{oil} + \text{salt}) + \text{heat}(\text{tomatoes} + \text{spices} - \text{water}) + \text{cheese} + y] = \text{pizza}$$

Of course, y = toppings, so if you solve for y, you'll know what toppings you want.

Keep in mind the distributive property, which states that if the equation is balanced, cheese, sauce, and toppings will be evenly distributed across slices, though not necessarily across bites. Meanwhile the associative property states that if people serve you subpar pizza, you will lose the desire to associate with them.

ALGEBRAIC COBB SALAD

To illustrate that even foods with many ingredients can be reduced to an equation, I present this recipe for a Cobb salad:

$$\left[\frac{\text{heat}(\text{egg} + \text{water}) + \text{heat}(\text{bacon} \times \text{smoke}) + (\text{avocado} \div \text{knife}) + (\text{tomato} \div \text{knife}) + (\text{chicken} \times \text{grill}) + (\text{Roquefort} \div \text{fingers})}{\text{romaine} + \text{endives} / \text{knife}} \right] \times \text{red wine vinaigrette}$$

See, isn't that so much simpler than a traditional recipe?

USING LOGIC TO PROVE THAT BACON IS OVERRATED

Eaters aren't afraid to challenge peoples' assumptions about consumption. So it is with sound mind and palate, and full understanding of the backlash it will create, that I state the following: Bacon is overrated.

That is not to say it isn't great. Bacon is great. But it's not as great as the Internet would have you believe, and it has become an overused meat stuff.

I have arrived at this controversial theorem using inductive logic:

- If a food has become so pervasive that entire websites compete to find the most absurd method for its inclusion—see bacon gumballs, bacon soap—then that food and its powers are rated very highly and it is widely used.

- If a food has an intensely salty and smoky flavor, it overpowers many of its culinary compatriots. Therefore its use should be limited.

So bacon is highly rated and widely used, even though its flavor is overpowering and, as pork products go, one-dimensional. Thus, it is rated more highly than it should be, meaning it is overrated.

Put another way, bacon is like Al Pacino. It's very talented. When you want someone to yell a lot, there's nobody better. But when overused it drowns out other performances, and each year it has to yell a little louder to have the same impact. That's why it's so tough to pair it with the right costar, and why Chris O'Donnell is known as "the Tomato of Hollywood."

HOMEWORK

Some people love math because there are clear right and wrong answers. Others hate it for the same reason. But when you study it at a very high level, as we have, you find that there's plenty of art involved in the science of crunching numbers and snacks. Conceiving delicious new math theories requires just as much creativity as composing great symphonies.

For your assignment, try to solve the Incompatible Food Triad. It's not clear who first posed this question, although the Internet's best guess is a philosopher named Wilfrid Sellars. (It's been popularized online by the mathematical sculptor and designer George W. Hart.) The IFT has been around for decades, and I'm always curious to hear what other Eaters come up with. So here it is, as posed by Hart:

> *Can you find three foods such that all three do not go together (by any reasonable definition of foods "going together") but every pair of them does go together?*

These foods that "go together" can be ingredients, seasonings, toppings, or separate whole foods that you combine into one dish and eat as one. A sample solution would be three nacho toppings—A, B, and C—where nachos topped with A and B are good, nachos topped with A and C are good, and nachos topped with B and C are good, but nachos with A, B, and C are bad. Good luck!

Submit your homework to me at dan@sporkful.com.

8

PSYCHOLOGY
Finding Yourself, Finding Your Food

- The stomach as psyche

- Early childhood Kit Kat trauma and Piaget's theories of development

- Object permanence and Refrigerator Blindness

- Oral fixations and better pen chewing

- Maslow's Hierarchy of Needs

- Relationships

- Some foods just aren't that into other foods

- Playing the field with per-bite dipping

- Dunking positions to satisfy both partners

- Dunktrimonial bliss with the Sidecar

Tolstoy wrote, "Happy families are all alike; every unhappy family is unhappy in its own way."

What Tolstoy failed to recognize is that every hungry family is unhappy, which means that hungry families are all unhappy in the same way. Also, some happy families are happy because they have pepperoni on their pizza. Others are happy because they don't. Totally different.

Of course, true happiness isn't as simple as a two-thousand-page Russian novel makes it sound. It requires self-acceptance, self-love, and ultimately, self-actualization.

In his seminal work *Civilization and Its Discontents*, Freud essentially argues that the pleasure principle drives Eaters to seek deliciousness, but that civilization's rules and restrictions stymie our efforts and lead us to feel discontent.

He's right. But where he and I differ is that he considers this to be an unavoidable side effect of organized society. I believe it is something that we can overcome. Many of civilization's "rules" are mere guidelines that can be cast aside, such as the idea that adults shouldn't trick-or-treat or that appetizers must be consumed before entrees.

Other rules are truly insurmountable, in which case we Eaters have a responsibility to seek deliciousness by working within their confines—for instance, by understanding the difference between dipping and dunking and employing each technique appropriately. In so doing, we reduce resentment and increase happiness across the Eatscape. This is the definition of Eater Actualization at Sporkful University.

In this chapter we'll discuss the path to that summit of the psyche, sometimes studying it from the perspective of the Eater, other times from the perspective of a food. Like people, foods interact with and influence each other. They have their own desires and goals, triumphs and disappointments and dysfunctions. And like us, their journey of self-discovery begins at birth, continues a lifetime, and is perhaps best summed up by psychologist Abraham Maslow's Hierarchy of Needs, which starts with the basic necessities of life and works its way up to the components required for self-actualization.

Along the way you'll learn how to relate to yourself and the Eaters and foods around you in a more constructive way, and gain a deeper understanding of how foods should be combined and consumed. These are crucial skills. After all, how will you ever find Perfect Deliciousness if you can't even find yourself? And how will cherished foods accompany you on the journey if they're as lost as you?

CHILD DEVELOPMENT

Freud observed that the experiences of early childhood have a lifelong impact on the psyche, which is why adults confronted with marinara, masala, or moonshine that's not how Mom used to make it experience mental anguish at holidays like Thanksgiving. This phenomenon is known as the Psycho-Gravitational Pull of Oedipal Stuffing, an issue I'll cover in more detail in my next book, *Turkey from the Teat*.

The most important thing for you to know about child development is that at any moment, with one false move or bad parenting decision, you could screw up your kid for life. Act wisely at all times, and you'll set your offspring on the path to Eater Actualization.

A CASE STUDY IN TRAUMATIC CHILDHOOD CANDY CONSUMPTION

Let's begin with a cautionary tale. When children experience trauma relating to something they hold dear, the wound cuts deep. That's why parents must be especially careful when administering candy, one of the greatest childhood treasures of all.

Here's a case study to illustrate this point . . .

Eater Patti Woods of Trumbull, Connecticut, is now over forty years old, but ever since she was a child, she's loved Kit Kats.

She's put a great deal of thought into the ideal way to consume her beloved treat, finally settling on a method whereby she breaks off one bar—correctly called a finger—and eats off all the exterior chocolate, corn-on-the-cob style. She then turns the finger on its side, so the wafer and chocolate layers are vertical, and nibbles each layer individually, to prolong the experience (figure 8.1).

Eater Patti's "Working Through Some Issues" Approach to Kit Kat Consumption

After you've eaten off the exterior chocolate corn-on-the-cob style, turn the finger so the layers are vertical and eat them one-by-one.

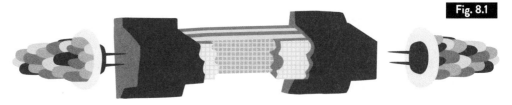

Fig. 8.1

Only someone who has endured Kit Kat–related trauma at a tender age could invent such a technique. That's why I was not surprised to learn that when Patti was five, she was with her mother in line at the supermarket. There was a boy in the cart ahead of them, and that boy's mother gave him a classic four-bar Kit Kat. He unwrapped the paper and the silver liner, but rather than break off a single bar, as Patti had been taught to do, he bit into the entire structure at once, across perforations, with total disregard for decorum and the Order of Things.

This happened nearly four decades ago, but Patti still has not recovered. On *The Sporkful*, she told me, "I wish I could find this boy, who I assume is now a man, and tell him that this was so wrong."

I'm skeptical as to whether he ever made it to adulthood, considering the savagery of his upbringing.

Fortunately, Patti is breaking the cycle. She had a child, she says, so she'd still have an excuse to go trick-or-treating. And each time her daughter gets a Kit Kat, you can be sure Patti is there to guide her.

JEAN PIAGET ON CHILD DEVELOPMENT AND CHOCOLATE

In order to better understand Patti's story, let's turn to Jean Piaget. He was a Swiss developmental psychologist, so he knew a thing or two about chocolate and children. Piaget identified four key concepts to explain how children learn. Here's each one and how it applies to candy:

- A **SCHEMA** is a child's understanding of a candy and its preferred method of consumption. (Example: "A candy bar is a delicious block of chocolate. I put it in my mouth and eat it.")

- When children add new information to an existing schema, that's **ASSIMILATION**. (Example: "Kit Kats have wafers inside, and they are also candy bars. Mommy breaks a piece of the Kit Kat off with her hand and eats it. I'm going to add this information to my candy bar schema.")

- When new information causes children to change a schema, that's **ACCOMMODATION**. (Example: "That boy just bit straight into a Kit Kat bar, across the perforations. The world is a sad, depraved place.")

- The natural attempt to balance assimilation and accommodation is **EQUILIBRATION**. (Example: "I'm going to continue breaking

TRICK-OR-TREATING FOR GROWN-UPS

If you have children, you can easily dip into their Halloween candy. But if you don't, you may come to resent society's rule that trick-or-treating is for kids. When that happens, get creative to score your own loot. Here are some suggestions:

1. Have kids. Plenty of people have been brought into the world for less noble reasons.

2. Dress up your pets. My aunt Meryl is a fiftysomething-year-old woman who thinks and eats like an eight-year-old. She has no kids but lots of dogs, and she loves candy. So each year she dresses up her pit bull, Katie, like a princess, hangs a bag around Katie's neck, and heads out into the neighborhood looking for loot. You'd be amazed at how much candy you can get when you trick-or-treat with a pit bull (figure 8.2).

Fig. 8.2

3. Have no shame. Meryl explains her strategy: "I take Katie to the neighbors and say, 'So, what are you giving out this year? Do you want Katie and me to do a quality-control check?' They usually feel so sad for us that they give us some candy."

Unfortunately, Meryl's haul diminished recently after Katie had surgery to repair a nagging limp, making them both appear less sympathetic.

Pit Bull as Adult Trick-or-Treating Companion

off Kit Kat fingers by hand, while at the same time taking a solemn vow to rid the world of candy bar ignorance.")

Now let's apply Piaget's theory to the boy in Patti's story and see how parental negligence can derail a fragile child. Here's the same encounter from his perspective:

- **SCHEMA** "A candy bar is a delicious block of chocolate. I put it in my mouth and eat it."

- **ASSIMILATION** "Kit Kats have wafers inside, and they are also candy bars. Mommy says I should stop asking how to eat a Kit Kat and just be quiet for one goddamn minute. I'm going to add this information to my candy bar schema."

- **ACCOMMODATION** "That girl who's watching me eat my Kit Kat just looked at me like I decapitated her doll. She must hate me, which probably means I should hate myself."

- **EQUILIBRATION** "I'm going to continue biting straight into Kit Kat bars, and also start biting into humans."

So you see, parents, you must always be aware of what you say and do around your children. If they end up being Kit Kat–defiling cannibals, it's most likely your fault.

ORAL FIXATION AND PEN CHEWING

Freud identified the oral stage as the first of five stages of psychosexual development, because an infant derives comfort and pleasure from eating and sucking. When the child is weaned from the breast it may cause anxiety that continues into adulthood, in the form of an oral fixation.

If that sounds like you, don't blame your mother. Just find a healthy outlet for it, in the form of a pen. Of course, for your pen-chewing habit to be considered healthy, you must chew on your pens properly, which first means not actually digesting them. But it means so much more.

From a chewability perspective, there are two basic types of pens: capped and retractable. Capped pens have a cap to protect the tip. Retractable pens have a push button at the back end that pulls the tip inside the body. The latter are more common these days for writing, but they're vastly inferior for chewing.

Here are the problems with retractables:

- They have no caps (which are delicious on their own).

- Their pocket clips are usually fastened tight to the pen body and thus less chewable. The clips also make the surface of the pen body uneven, so you can't get a good lip seal, which is crucial to generating the suction needed for saliva containment.

- The click buttons render the butt end cumbersome and unsatisfying, and if you chew the end too hard and break the push-button mechanism, the pen is shot.

THE CAPPED PEN: SUCCUBUS OF SCRIPT

While retractable pens are bony, capped pens are buxom. They offer three distinct erogenous zones—the cap, the clip, and the butt—and sometimes include an end plug that offers a fourth pleasure zone. The best capped pens allow for long and varied chewing to satisfy even the most orally fixated. There are just too many toothsinkable positions for me to list them all, but I will share a few of my personal favorites . . .

Sometimes a Pen Cap Is Just a Cigar. Or a Pipe. Or a Saxophone.
This series of maneuvers is ideal for the creative thinker. Once you've masticated the clip until it's bent all the way back into a U shape, you can smoke, suck, or blow until your mouth's content. When generating new ideas,

chew the cap like a cigar. When pondering your ideas, puff on it like a pipe. When celebrating good ideas, play it like a saxophone.

Cigar   Pipe   Saxophone

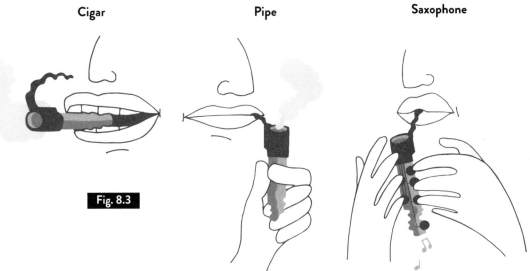

Fig. 8.3

Popping the Plug

This one only works when there's a removable plug in the butt of the pen. Chew the end until the plug comes out, then insert and remove the plug several times with your mouth. Set the plug aside and chew the butt of the pen for an extended period, until its opening is stretched. Finally, put the plug back into your mouth and chew it until it's folded in half, then reinsert it into the butt. Continue to use your mouth to remove and reinsert the plug as desired. When the pen has given you all it's got, you'll understand why Eaters call capped pens the porterhouses of penmanship.

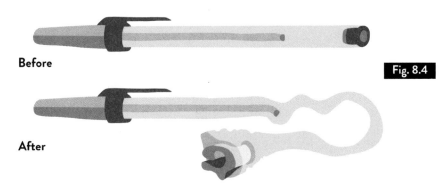

Before

Fig. 8.4

After

ALCOHOL AND CHILDREN: PERFECT TOGETHER

Lest this section leave you hopelessly resigned to turning your children into cannibals or pen perverts, let me offer a tip for child-rearing success.

My research shows that moderate alcohol consumption can reduce the impatience that often leads to bad parenting behavior. That's why, where groups of children are gathering with their parents, alcohol should be served—only to the parents, but for the benefit of all present.

The parents, of course, will welcome the many blessings showered upon the drinker, known to humankind for millennia, which include a general sense of well-being and a tendency toward good cheer. But what child would not also prefer pleasant, relaxed parents to stressed ones? And what child would not benefit from rearing by the former as opposed to the latter?

Alcohol's role in effective parenting, however, goes far beyond providing liquid patience. That's because when your children are young, you are the all-powerful creator of their social circle.

> TIP In college: Alcohol = Liquid Courage.
> In parenthood: Alcohol = Liquid Patience.

They don't really choose their friends. You choose *your* friends, then tell your kids to play with their kids. (When children get older, this equation is often reversed.)

Who, then, to befriend? And how? Birthday parties and other large gatherings with many small children can be overwhelming. The grown-ups rarely have more than thirty consecutive seconds with which to speak to each other, which makes the formation of true friendships difficult. For parents, children's birthday parties are like speed dating without the sex.

Enter alcohol. This timeless social lubricant allows you to skip the formalities and engage in meaningful conversation more efficiently. Plus, it helps you determine who else shares your superior approach to parenting, so you're able to make a good friendship match for yourself and your children at the same time.

In short, the next time you're hosting a large gathering with lots of kids, please ensure there is alcohol for the adults, regardless of the time of day. It makes the party better for everyone.

OBJECT PERMANENCE AND REFRIGERATOR BLINDNESS

Understanding object permanence—the idea that an object continues to exist even when you can't see it—is a crucial step in an infant's psychological development. But long after a child surpasses this milestone, blind spots may remain.

For instance, when my beloved Mrs. Sporkful scans the refrigerator, she often has no recollection of the existence of certain foodstuffs that are now out of sight, many of which she placed there herself just hours earlier.

Like so many, Mrs. Sporkful is afflicted with Refrigerator Blindness Syndrome.

If you live with someone who has this condition, you know it has its pros and cons. When you take care to leave a special dish in the fridge for the RBS sufferer, only to have it inadvertently wind up behind the milk, your efforts are squandered. When you're counting on this person to help you finish leftovers before they spoil and said leftovers end up under the cheese, you're on your own.

But when you wish to sequester a prized morsel for your mouth only, it's as easy as taking short ribs from a baby. And while some say it's wrong to take advantage of a loved one's disability, my research shows that what they don't know they didn't eat won't hurt them.

THE HIERARCHY OF NEEDS

Let's transition from looking at childhood to looking at the entire arc of our existence, so we can delve into the personal and psychological growth required to attain fulfillment in your soul and stomach. The best way to understand this growth is through psychologist Abraham Maslow's Hierarchy of Needs, which states that as more basic necessities are met, we can turn our attention to loftier needs. The hierarchy is often depicted as a pyramid:

Maslow's Hierarchy of Needs

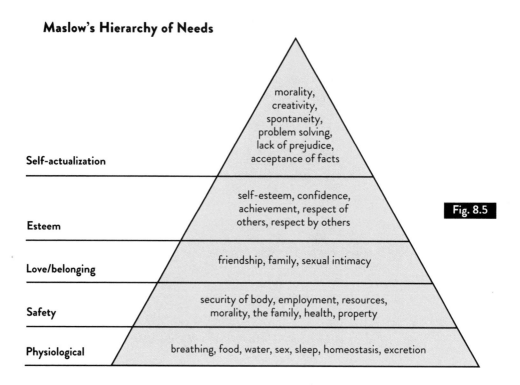

Fig. 8.5

It's pretty clear how the Hierarchy of Needs applies to an Eater, but to illustrate how it applies to a food, let's consider the cheeseburger (figure 8.6). Its physiological needs—ground beef, a bun, and cheese—must be met before it can develop the self-esteem necessary to consider relationships with toppings or condiments. If a self-actualized burger chooses to welcome such additions, it is not out of feelings of desperation or inadequacy, but rather out of confidence and a desire for pleasure on its own terms.

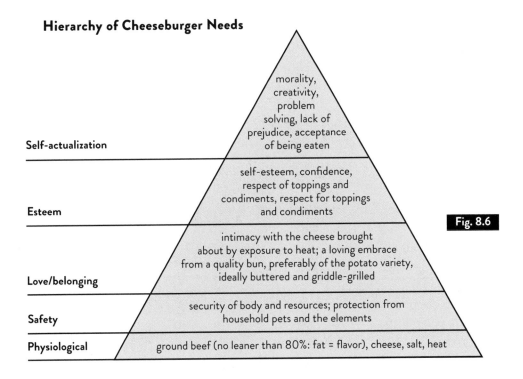

Hierarchy of Cheeseburger Needs

Self-actualization — morality, creativity, problem solving, lack of prejudice, acceptance of being eaten

Esteem — self-esteem, confidence, respect of toppings and condiments, respect for toppings and condiments

Fig. 8.6

Love/belonging — intimacy with the cheese brought about by exposure to heat; a loving embrace from a quality bun, preferably of the potato variety, ideally buttered and griddle-grilled

Safety — security of body and resources; protection from household pets and the elements

Physiological — ground beef (no leaner than 80%: fat = flavor), cheese, salt, heat

Now we'll work our way up Maslow's hierarchy, learning how to satisfy needs at each level, as we and our foods climb together to the acme of actualization.

THE THREE PARTS OF THE STOMACH: ID, EGO, AND SUPEREGO

We'll begin with the first two tiers of the hierarchy—Physiological and Safety. To understand the Eater's physiological needs, let's look at the three parts of the stomach: the id, the ego, and the superego. They are defined as follows:

- The **ID** is the part of the stomach that growls. It wants to simultaneously consume every delicious meal you've ever eaten, all at once, with cheese melted on top.

- The **SUPEREGO** is the part of the stomach that feels revulsion at the thought of doing that and produces pangs of guilt (also called "nausea") if you actually do it anyway.

- The **EGO** is in between the two, which is where the stomach's fill line is located, allowing the ego to have a realistic idea of how much food should be consumed at a given time. But it often rationalizes increased consumption, saying things like "You know, it doesn't really make sense to stop eating with so much empty space still in here. We really should fill that space with food, you know, to be more efficient. And for the environment. And the starving children."

> **TIP** A Freudian slip is when your conscious stomach tells you to eat some broccoli, but your utensil instead stabs a pork rind.

WHAT COLAS WANT

They may look different on the outside, but on the inside, colas all have the same physiological and safety needs:

- A cold storage environment so that ice is optional, not required

- Ice added only to dilute excessively sweet cola as desired, not to chill it

- Cold water as a dilution alternative to ice, if the Eater deems evolving levels of dilution to be a negative (it's essentially like adding ice that's not so cold)

- A pleasant release from cans and bottles via the spaciousness of a glass, which relieves it of excess fizz, or fountain service with correct syrup-to-soda ratios (increasingly a mythological scenario)

- Straw use only when ice is present, to reduce tooth clinkage

LOVE AND RELATIONSHIPS

Let's move up to the third and fourth tiers of needs, Love/Belonging and Esteem. At the core of Maslow's hierarchy is the idea that you'll never be truly happy unless you love yourself, love others, and are loved by others. Friends, family, and significant others satisfy those needs, but you're more capable of making relationships meaningful when you've also satisfied the needs for self-esteem and confidence.

FIRST-DATE EATING: WHAT YOUR SHARING POLICY SAYS ABOUT YOU

A person's approach to sharing food with others is a great way to determine how far up Maslow's hierarchy they've climbed and how ready they are for a real relationship. True Eaters share food readily, but the less assured are hesitant. Here's what you can tell about your date based on his/her method of sharing:

- **DOESN'T SHARE FOOD AT ALL** This person has been hurt before and now has a real fear of commitment, coupled with anal-retentive tendencies. Even after you break down their culinary barriers and get a little nibble, other issues will linger.

- **CUTS OFF A BITE AND PLACES IT ON YOUR PLATE** This person wants to share new experiences but also places a high priority on boundaries. This taste is a nice tease, but it may be a while before you're swapping fork spit.

- **HANDS YOU THEIR FORK WITH THE BITE STILL ON IT** This person eschews pretense, knows what they like, and has the confidence to go for it. Savor that bite, and look forward to seconds.

- **HANDS YOU THEIR ENTIRE PLATE AND LETS YOU TAKE WHAT YOU WANT** Tough call. This person could be a supportive and unselfish partner who wants you to choose your own path, or a lazy one who's unprepared for the hard work of a real relationship. Take an open-minded mouthful, but chew with caution.

- **STICKS THEIR KNIFE AND FORK INTO YOUR PLATE AND TAKES A BITE WITHOUT ASKING** This person is an asshole.

- **FEEDS YOU THE BITE FROM THEIR FORK** Ooh la la! Some may find this a little forward, but don't assume that anyone who's this into you has something wrong with them. Sink your teeth into this bad boy or girl.

Keep these guidelines in mind, and you'll never waste a second date again!

WHERE TO EAT ON A FIRST DATE

Not all great relationships begin with a great first date—but it sure helps. Eater Elizabeth Chubbuck is a cheesemonger at Murray's Cheese in New York, and she's been on more first dates in the last few years than I have (which means more than zero). Here's her take on finding the right restaurant and the right approach:

"You want a place that's unpretentious and unfussy, where the menu is not too extensive. You don't want to be bogged down reading the menu . . . Any guy that demonstrates any sort of restraint or concerns about his waistline or the health qualities of the food is not going to make the cut. Food and eating is all about pleasure and indulgence.

"My personal ordering strategy is to go for exactly what I want. That's also how I operate on the first date. If I want the beef heart that's been marinated and stewed for five days, then that's what I'm going to order, and if it makes him sick, then we probably shouldn't go on a second date."

Sorry, fellas, but I'm not at liberty to disclose Elizabeth's contact information in this book.

DIP ME FIRST, DUNK ME LATER

Just like people, foods fit together in different ways and want different things. We can learn a lot about a food by studying the types of relationships it seeks and forms, and more importantly, with whom it chooses to spend eternity.

In other words, how does a food want to be eaten?

Dipping and dunking are beautiful and sensuous ways to bring foods together, but these terms are not interchangeable.

Dunking involves submersion with an intent to bring about absorption, such that the food that is dunked is forever altered—like after a night of passionate lovemaking and/or electroshock therapy. "Dipping" refers to a more delicate kiss. It's the petting to dunking's impregnation, and while it can also denote penetration, its effects are mostly reversible. (In other words, you can wipe ranch dressing off a carrot, but you can't wring coffee out of a cruller.)

Foods should not join together in dunktrimony unless they're really serious about each other. Successful dunking requires both foods to be self-actualized, so we'll discuss that more in the next section. For now let's focus on dipping, a better option for foods that have found love and esteem, but haven't fully found themselves.

PLAY THE FIELD: DIPPING KEEPS YOUR OPTIONS OPEN

Too often, Eaters force foods into arranged marriages through spreading or dunking. When you spread any condiment or sauce on an entire food at once, you eliminate an array of eating options in one ill-considered stroke.

Want to try your cheeseburger with a different condiment, a different quantity of condiment, or no condiment at all? *Too late.* Sorry that that chili is turning your nachos to mush? *Your bad.* Concerned that the sauce in your chicken parm sub is soggifying the roll and imperiling its hinge? *Wish you'd have dipped.*

Dipping each bite into a condiment, spread, or sauce opens up possibilities and eliminates pitfalls (figure 8.7). It allows you to reduce soggage, alter ratios on the fly, and create dozens of variations on the food before you.

Most importantly, it offers you and your foods an opportunity to play the field at a time when experimentation and self-discovery are crucial.

Techniques for Dipping Sandwiches or Burgers on a Per-Bite Basis

It's up to you to decide exactly how much dip you want in each bite, but in general, remember that the vast majority of condiments and sauces are meant to be supporting actors, not leading players. The easiest way to moderate a dip's influence is to consider how much of the bite surface—the area where you plan to chomp—actually touches the dip. Here's the wrong way: - - - - - - - - - - - - - →

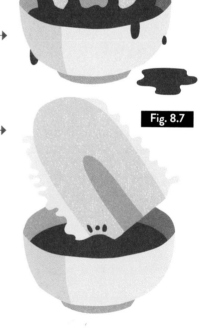

Fig. 8.7

See how the entire bite surface of the above sub half is touching the dip? What is this, a shotgun wedding? Slow down! Just dip a corner, to add a more subtle accompaniment: - - - - - - - - - - - - - →

Dipping an unbitten burger is tougher because it's round and has no points or corners to dip. That's why I like my first bite of burger uncondimented. Not only does it create a crescent-shaped indentation with dippable points, but it also lets me gauge what condiments are needed, if any, which is just the flexibility I seek.

BULLYING AND ALTERNATIVE-LIFESTYLE FRENCH FRY DIPS

Gaining the respect of others—level four in Maslow's hierarchy—can be difficult if your lifestyle is considered "outside the norm." Foods that are different are often targets of bullying.

For instance, some say French fries should only be dipped into ketchup. They even think that a basic alternative like mayo is limp-wristed and European. It's up to us, the bystanders of the Eatscape, to stand up against such closed-mindedness. Here are some dip partners that French fries choose when they have the freedom to eschew more conventional ways:

BUTTER Make sure it's at room temperature, or even a little warm. Salt the butter and the fries, but don't go overboard on either. Season the butter with hot sauce, honey, or whatever else strikes your fancy.

GUACAMOLE It's not just for tortilla chips. A thick, crispy, properly salted fry can handle a dollop of guacamole, and they'll both be better for it.

GRAVY Gravy slovenly smothers *poutine* and disco fries, but when given the chance, it cleans up real nice. It's always been a delicate dip trapped inside a tawdry topping's vessel. The core of gravy's identity is in its meat juices—not in butter, cream, or flour. Use a gravy that knows itself well, and you shall have a fast fry friend indeed.

MASHED POTATOES French fries love the sweet caress of their creamy cousin, and they'll thank you not to judge them for it. (There's a reason why Napoléon dropped incest from the French penal code, a decision that inspired the creation of the French Fry Napoléon—see page 232.) Sprinkle the mashed potatoes with fried onions to add more crunch.

MARINARA SAUCE This works along the same lines as ketchup but with more focus on pure tomato flavor, less on sugar. Use a thicker marinara so it stays on the fries.

FRENCH FRY NAPOLÉON

Your relatives can drive you crazy, but they're also the people who know you best. That's why these cousins from the potato family have such a strong bond. Use wide, flat steak fries, which should make it easier to pick up the structure and eat it in one or two bites. (Cutting is likely to force mashed potato out the back.)

Fried onions add crisp and flavor

Steak fries

Mashed potatoes

WHY CHIP-AND-DIP MAKES A BETTER RELATIONSHIP THAN CHIP-AND-DUNK

Respecting the line between dipping and dunking sets useful boundaries in a relationship between foods. It also has practical advantages for the Eater. If you mistakenly use a dunk technique when you mean to dip, you put undue pressure on the foods' relationship and increase the chance of a chip breakup.

Remember that one of the key differences between dipping and dunking is that dunking implies the long-term commitment of submersion, while dipping calls for a lighter touch. When dipping a chip, fry, or almost anything else, pay close attention to the angle from which you enter the dip to ensure you don't accidentally cross over to a dunk. Walking this line can make the difference between a relationship that starts slow and lasts, and one that speeds ahead, only to end up crushed beneath the weight of expectations and guacamole.

In other words, don't plunge the chip straight down into the dip roughly, like a pubescent schoolboy discovering a young girl's salsa garden at summer camp. This puts more pressure on the chip and increases the chance it will break up, an inevitable ending for summer camp romances.

Instead, sweep gently and horizontally across the top of the dip, as you'd caress a lover's bare hip with your fingertips. By taking it slowly, you put less pressure on the relationship and increase the chance that it will be built to last, all the way to your mouth.

SAUSAGE GRAVY CONSUMPTION AND
FEELINGS OF SELF-WORTH OVER TIME

Some behaviors bring you closer to achieving self-love. Others push you further away. Research shows that life experiences provide more lasting happiness than material possessions, which generally offer only short-lived pleasure.

This is the problem with sausage gravy. It's so thick and heavy that its consumption is not just a life experience, but also the acquisition of a material possession—in your stomach. Like that new dress or tech gadget, you think it'll make you happy, but it won't. Instead, it weighs down the psyche and the stomach, reducing the Eater's self-esteem.

This chart explains the phenomenon:

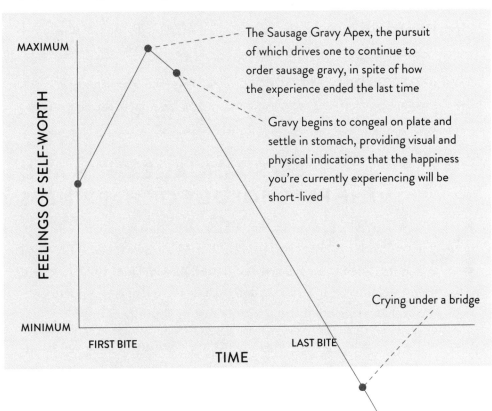

PSYCHOLOGICAL DISORDERS

Before you can proceed to self-actualization, you'll need to determine all the ways in which you are mentally ill. Here are some possibilities:

GENERALIZED ANXIETY DISORDER You constantly worry that your food won't be delicious enough and/or that there won't be enough of it.

OBSESSIVE-COMPULSIVE DISORDER You compulsively turn on the stove to cook things, taste food repeatedly, and wash your hands after eating.

PANIC DISORDER You suffer from recurrent panic attacks, perhaps resulting from agoraphobia (the fear of being in a public place without enough food) or claustrophobia (the fear of being in a confined space without enough food).

HYPOCHONDRIA You always think you're hungry, even when you're not.

DEPRESSION After an especially great meal is over, you become deeply sad and withdrawn.

MULTIPLE PERSONALITY DISORDER When free food is available, you assume multiple personalities to take as much as possible without seeming greedy.

SELF-ACTUALIZATION AND THE PSYCHOLOGY OF HAPPINESS

Now we arrive at the pinnacle of Maslow's Hierarchy of Needs, self-actualization. Of course, getting here requires much more than reading this far. I expect this chapter, like much of this book, will only be the beginning of an extended period of soul-searching and self-exploration that will motivate you for the rest of your life.

One indicator of self-actualization in foods is the ability to form deep, lasting relationships consummated with a dunk. We'll begin there, then move on to other behaviors associated with Eater Actualization.

DUNKING COOKIES AND DONUTS: SETTING CLEAR BOUNDARIES WITH THE SIDECAR

The two partners in a dunk relationship are the alpha and the beta, the alpha being the food that is dunked, the beta being the food that receives the alpha. The transference of liquids here is irreversible and constitutes a lifelong commitment between the foods.

That transference can be mutual or unrequited, good or bad. When you dunk a chocolate chip cookie into milk, some cookie also ends up in the milk. Were it enough to create some sort of delicious cookie milk, you could argue this was advantageous. But it's usually just enough to add an unpleasant texture to the milk, without really changing its flavor.

Dunking a donut into coffee is even more problematic, as donut pieces are more easily dislodged from the whole, and turn to mush even more quickly when left in liquid. If you've ever dated someone who took a piece of you and destroyed it while at the same time destroying themselves, you don't need a shrink to know that this is not the basis for a good relationship.

The best way to address issues like these is to establish clear boundaries. I've helped many dunking partners do just that with the use of a sidecar (figure 8.8). Set some milk or coffee aside in a teacup or espresso cup, and dunk the alpha in the sidecar to keep your primary beverage unadulterated.

Dunking with the Sidecar

Fig. 8.8

Crumbs remain isolated in the sidecar,
leaving primary coffee pure.

WHEN YOU'RE READY: DUNKING POSITIONS
TO SATISFY BOTH PARTNERS

When foods come together in holy dunktrimony, there's a lot of focus on the nuptial dunk. But the truth is, that first time is often clumsy and unwieldy. Large, rounded alphas like cookies and donuts may not fit easily into the narrow opening of a sidecar, and overzealous Eaters must remember that force is never the answer.

Breaking cookies and donuts before dunking not only transforms them into more dunkable shapes but also exposes their interior, which pleases the milk and coffee, who want their partners to open up and let them in. After all, these beverages aren't looking for a one-night dunk. As coffee cools and milk moves toward curdling, these once-libertine libations are reminded that their biochemical clocks are ticking.

GRILLED CHEESE AND TOMATO SOUP:
A RELATIONSHIP READY FOR REPRODUCTION

Anyone who has ever gotten married knows that the first question people ask you the day after your wedding is "When are you going to have kids?" Why should the question we ask a classic dunking duo be any different?

Grilled cheese and tomato soup have been a rock-solid pairing for decades, so it's surprising that they have no offspring. (I'd love to host a family gathering where all sorts of little grilled cheese sandwich–and–soup combinations scamper around the table and into my stomach.)

Perhaps grilled cheese and tomato soup have been trying to get pregnant for a while, but they're having trouble. In that case, members of the Eatscape must use our kitchen labs as fertility clinics.

As we do, let's remember what makes this relationship so special. While more of the transference moves in the direction of the sandwich, the melted cheese and buttery bread lend the soup enough richness to make this an instance of mutual transference that's beneficial to all.

Respect, communication, give-and-take—this is the foundation of a lifelong partnership and the recipe for dunktrimonial bliss. Now let's see if we can replicate it.

GRUYÈRE GRILLED CHEESE WITH CREAM OF MUSHROOM SOUP

For this combination I recommend a thin, crusty bread, which will stand up to the weight of the soup without adding a lot of mass to an already rich combo.

RICOTTA, PARMESAN, AND SPINACH GRILLED CHEESE WITH SLOVAK GARLIC SOUP

Slovak garlic soup is usually made with a stock base and no cream, so it's light in color and consistency. The key here is a softer bread, as a bread that requires a lot of bite force will cause ricotta to fall out the back.

GRILLED CHEESE ENCASED IN CRISPY CHEESE WITH CHEESE FONDUE

Prepare cheese fondue and set aside in a fondue pot, or other warm vessel. Cook the grilled cheese sandwich of your choice in a nonstick pan. When it's finished, lift it out of the pan, place 1 quarter-inch thick slice of provolone into the pan, put the grilled cheese on top of it, and place the other slice of provolone on top of the sandwich. Adjust the burner to medium heat.

Let the sandwich cook for 8 to 10 minutes or until the underside of the provolone is an even dark golden brown. Slide a rubber spatula underneath periodically to prevent sticking. Flip the sandwich and cook it until the second side is an even dark golden brown. Remove the sandwich to a cutting board, slice it into four wedges, and dip it into the cheese fondue. Die.

Ice Pop Cocktail Stirrers: Mutual Transference Mastered

Here's another example of a successful dunking relationship marked by mutual transference. This is a great way to sweeten a summer cocktail and add some zing to an ice pop at the same time. If you monitor pop meltage carefully and alternate between drinking and licking, you'll improve both your beverage and your pop.

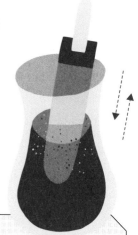

SOME FOODS JUST AREN'T THAT INTO OTHER FOODS

Our societal focus on dipping and dunking puts a lot of pressure on foods that struggle to find the right partner. Some assume that if a food doesn't pair up, it must have something *wrong* with it. As these foods approach their expiration dates, that pressure only mounts, and even well-intentioned Eaters can become overzealous in their attempts to fix up friends in the fridge.

Consider the case study of pretzels. The difference between a hot, fresh, oven-baked pretzel and any other hot, fresh, oven-baked bread product is that the pretzel tastes like pretzel. Dunking it in anything that overshadows this flavor is a rejection of its very identity. This applies not only to the cheese-based sludge that so many bars provide, but also to pretzel's longtime companion mustard.

Indeed, fresh pretzel has had many dipping dalliances with mustard over the years, but I've advised it against the long-term commitment inherent in dunking. Anything more than the slightest kiss renders the pretzel little more than a mustard delivery system and essentially says to the pretzel, "I love you, you're perfect, now change."

THE ACTUALIZED EATER

While self-actualization is generally a prerequisite for finding the right life partner (see previous discussion on dunking), a life partner is not a prerequisite for self-actualization. Maslow says certain behaviors are signs you've reached the summit of the self of the stomach: morality, creativity, spontaneity, problem solving, lack of prejudice, and acceptance of facts.

Here are some examples of those behaviors in action:

MORALITY: ON NACHO SELECTION

You can tell a lot about an Eater by watching how he or she behaves when a heaping pile of nachos shows up. That's because there's usually a top layer of melted cheese fusing several dozen tortilla chips together, and when everyone starts reaching for their share, some people may end up with much more cheese than others. Of course everyone's entitled to take one nacho, but what exactly constitutes "one nacho"? If twenty-five chips are fused together into one mass and you remove that entire layer, including all the cheese, did you just take one nacho? And is that moral?

Most of the time, there's nothing immoral about trying to assemble a great bite. But a scorched-earth approach to nacho consumption is not going to help you win over friends or foods. You're not only alienating fellow Eaters, you're also leaving lower-level chips without access to melted cheese, making those bites worse for everyone—including you.

The guiding principle here is called the One Hand, Two Chips Rule of Nacho Morality. It states that when taking nachos from a pile, you should only use one hand and only grab up to two chips at once with that hand. Whatever comes along for the ride is part of your nacho allotment, but this generally makes it difficult to remove an immoral proportion of melted cheese at once. If an Eater follows this rule and still ends up pillaging cheese resources, the nachos themselves are at fault.

CREATIVITY: THE BACONY BUNDLE OF TWIGS

Grilled vegetables are delicious, but asparagus and string beans are generally too thin to keep on the grill without special apparatus. Enter the Bacony Bundle of Twigs. You want enough bacon overlap so the bacon fuses to itself as it heats and holds the asparagus in place. Keep movement to a minimum so the fusion holds. If you have to add a toothpick or skewer for stability, so be it, but ideally, the bacon here is more than just a pretty taste. Grill on low to medium heat until bacon is crispy.

SPONTANEITY: DO IT NOW

Try any one of the techniques in this section right now.

PROBLEM SOLVING: TWO DIPS ON ONE CHIP WITH THE SALSA SLALOM

Problem: You want to eat a tortilla chip with both salsa and guacamole on it, but the two dips are in separate bowls, and you don't want to vitiate one with the other. Solution: the Salsa Slalom.

Dip one corner of the chip in the thicker dip, and the other in the thinner dip. Start with the thicker dip to ensure it stays on the chip as you slalom your way to satisfaction.

LACK OF PREJUDICE: ENTREES FIRST AND APPETIZERS FOR DESSERT

The typical order of a meal—appetizer, entrée, dessert—constitutes a preconceived notion that may reduce deliciousness. The self-actualized

Eater rejects the prejudiced idea that every course must stay in its place within the meal, just as we reject the idea that every food must stay in its place within the day (as discussed on page 99 in "Philosophy").

When you're especially focused on a certain entrée, you want to maximize your enjoyment of that entrée by being as hungry as possible when you eat it. An appetizer will only take up stomach space and blunt your appetite, lessening your enjoyment of the food you sought in the first place. That's why I've never found amuse-bouches very amusing.

When I go to the 2nd Ave Deli in New York, I start with a hot pastrami on rye, when my appetite is primed and my stomach is empty. If I'm still hungry, I'll top myself off with soup or an appetizer. But the pastrami is the motivation for my visit, so that's always the first thing I want in my mouth.

Appetizers also make for excellent desserts. Perhaps you're still a bit hungry, but you're not in the mood for anything sweet. Maybe you want a second helping of a starter that was finished too quickly. What's stopping you? After one memorable meal in Italy, as my friends ordered chocolate cake, I had shrimp risotto. That could be you.

ACCEPTANCE OF FACTS: SOGGY BREAD AS A STRENGTH

Eater Actualization doesn't just mean accepting the facts of life, it also means embracing them and using them to inspire us.

Grilled bread is ideally suited for dunking in soup and broth. It soaks up a lot of liquid and flavor while maintaining crisp and crust, and the char of the grill marks adds another component. But this combination comes with competing concerns. As your bread soaks up broth, it becomes more scrumptious and yet more likely to disintegrate.

You can fight the bread's natural tendency toward self-destruction by removing it from the soup before it's fully saturated, but now you're depriving yourself of broth-based deliciousness. So don't fight the sog—embrace it.

EMBRACE THE SOG CROSTINI

Start with 2 thick slices of bread that have been brushed with butter or olive oil and grilled. Submerge 1 slice into broth or sauce until it's completely saturated. Gently lift or spoon the soggy bread onto the dry bread and let sit for a minute, so some moisture seeps into the bottom bread and the two fuse to each other a bit. Pick it up and eat it.

I love the textural contrast of this dish, which gets you maximum saturation along with plenty of crisp.

HOMEWORK

The ways in which Eaters and foods behave and interact has tremendous influence on a meal. Self-actualized Eaters navigate these relationships with confidence and respect for themselves and others.

Getting to that point requires hard work and consistent examination of yourself, your food, and your food's self. When you reach that zenith of the mind, your tongue will know it. But eating experiences that go awry can lead to grave unrest in the unconscious mind.

Which brings us to your assignment . . .

I had a dream in which I had dinner with family and friends at a restaurant on an island. After dinner I was trying to row a boat from the island back to my house, when I became lost in foggy waters. After great struggle, I finally got to shore. When I returned home, I realized I had left my doggie bag at the restaurant on the island. I woke up furious. Interpret my dream.

Send your interpretations to me at dan@sporkful.com.
Include your own food dreams and your interpretations of them!

9

BIOLOGY & ECOLOGY
Eatscape as Ecosystem

- Buffalo wing consumption techniques to reduce meat and napkin waste

- Evolutionary eating advantages and spice-loving species

- Biodiversity and snack mix

- The Brazil Nut Effect

- The Sandwich Genome Project

- Dominant and recessive ingredients and the Punnett Square Sandwich

- Gastronomy taxonomy

- Classifying pasta shapes

- In vitro popcorn fertilization

- Drunken Salami

Take a moment to marvel at the natural balance of the world around us. It's an innate equilibrium, set in place by great forces over millennia, that weaves together plants and animals, minerals and elements, air and water, sun and moon. It evolves so slowly that we hardly notice. Each piece is dependent upon another, so one change leads seamlessly to the next, and it's often impossible to figure out where anything begins or ends. It's an eternal ripple effect, a condition that the biologist Elton John has dubbed "the Circle of Life."

As Eaters, we have a responsibility to study and preserve this natural balance, and to eat in harmony with it. The Eatscape is our ecosystem. Within it, we are the creators of the culinary world where our foods live and interact, where every meal offers us the chance to play gastronomic god.

In this chapter we'll study the interconnection between Eaters and foods, and work to translate that better understanding of biology into deliciousness. We'll reduce meat waste through more thoughtful Buffalo wing technique and improve snack mixes by focusing on biodiversity. We'll cover genetics to learn how our foods are related on a microscopic level, and taxonomy to learn how to categorize them, so that we can choose between dishes more effectively. Finally, we'll step into our kitchen labs to see how our newfound knowledge can lead us to make great discoveries.

ENVIRONMENT AND EVOLUTION

We Eaters are the caretakers of the Eatscape, Gaias of the gut. We are preservationists, even at the exact moment that we're consumers. We embrace this responsibility both to ensure that the larger Eatscape flourishes for future generations, and because it offers vital lessons for the smaller ecosystems found on our plates and in our kitchens every day.

CONSPICUOUS WING CONSUMPTION:
REDUCING MEAT WASTE WITH BETTER TECHNIQUE

Eating Buffalo wings—or any sauced chicken wings—with the right approach reduces both napkin and meat waste. It's better for the environment and better for you, because you'll get more meat off the bones with less work.

Chicken wings generally come in two shapes, which the wingerati refer to as the drumette and the flat, or paddle. (The drumette is the mini drumstick; the flat/paddle is the piece with two parallel bones inside.) Eating a drumette is mostly self-explanatory, but here are a couple of tips that apply for both parts:

- Hold the wings with your fingertips, using as little of the pads of your fingers as possible. This reduces hand mess.

- Bite with your teeth, keeping your lips out of the way. This reduces face mess and spice-induced lip burning.

When a group of amateurs is faced with a pile of wings, the drumettes are often in greater demand, because they're considered

TIP Drumettes have an average meat-to-bone ratio of 0.49, while flats have an average MTBR of 0.62, according to competitive eater Crazy Legs Conti. And no, there aren't supposed to be quotation marks around Crazy Legs. It's his legal name.

easier to consume. But Eaters know that the flats offer better ratios of skin to meat, fat to meat, and meat to bone, which means more crisp, more flavor, more meat for your money, and more better meat. Learn to navigate the flat's two interior bones, and you'll find it's no harder to eat, and far more delicious. Here are three approaches:

The Bone Splitter

This is how I usually eat wings. My technique is extremely messy, but it's delicious, it leaves no meat behind, and you can still reduce napkin waste by waiting until you're finished eating to bathe.

Step 1: Hold the wing so the side with the thick layer of skin is facing away from you. Dig the tip of your thumb into the groove in the middle of the joint at the thicker end of the flat until the joint splits.

Step 2: While using one hand to surround the meat and hold it in place, use your other hand to pull out the thinner of the two bones (the one toward the thinner side of the flat).

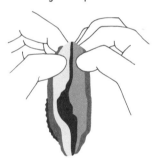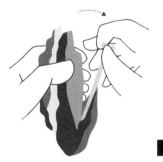

Fig. 9.1

You now have tender, succulent, toothsinkable meat with just one bone running through the middle of it. You can either pull the meat off the bone by hand, eat the wing like a rib, or stick the entire thing in your mouth and rake out the meat as in the Rake's Progress.

The Rake's Progress

I learned of this maneuver from competitive eater Tim "Eater X" Janus, who once ate more than five pounds of wings in twelve minutes. To keep up that pace you need a good strategy for the flats, and he has one. (For more from Eater X, see his Thanksgiving pep talk on page 156 in "Cultural Studies.")

Step 1: Hold each end of the flat with your fingertips and twist in opposite directions until one joint snaps.

Step 2: Spread the bones open into a V shape and insert the entire thing into your mouth. Use your teeth to rake the meat off as you remove the clean bones from your mouth.

The Meat Umbrella

I picked up this move from the aforementioned Crazy Legs Conti, who did it on my Cooking Channel web series *You're Eating It Wrong*. He uses one hand, but he has big hands. I use two.

Step 1: Stand the wing up on a plate or flat surface with the widest part of the wing at the bottom. Grip the wing firmly around the top and press in towards the bones.

Step 2: While continuing to press inward, move your fingers down the wing. The meat will strip off the bone as if you're opening an umbrella. Pop the meat into your mouth. It tastes so good, you'll be praying for rain!

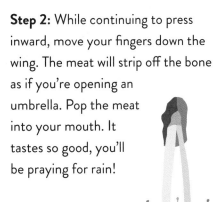

The Forked Tongue

My friend Brendan McDonald is a passionate Eater and past *Sporkful* guest, and he devised this technique for very spicy wings. I'm not a fan of extraordinarily hot wings, and spicy food in general is one area where Brendan and I disagree. (See also "On the Origin of Spices," below.)

Despite our differences, Brendan's approach is thoughtful enough to merit inclusion in the wing canon. I'll trust you to decide whether to use it.

Step 1: Nibble the flat's entire exterior, corn-on-the-cob style, making sure your lips don't touch the wing, to minimize spice pain.

Step 2: The remaining bit of meat between the two bones has not been sauced, so it's not spicy. Poke it out with your tongue and use it as a palate cleanser.

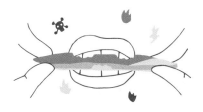

ON THE ORIGIN OF SPICES: A MORE EVOLVED THEORY ON BUFFALO WINGS

Some Eaters believe that Buffalo wings play a role in natural selection. As the logic goes, intensely spicy foods help weed out weaker organisms. Those who can withstand extreme heat survive and procreate, passing on their palates to the next generation.

Perhaps in some prehistoric era when physical strength was more useful than intelligence, high tolerance to hot foods helped you out. But Darwin's theory of evolution emphasizes the capacity of species to adapt to changing

WHEN ORDERING WINGS, RESPECT THE FRIED

With all the focus on the sauce, many Eaters forget that even though Buffalo wings are not breaded, they're still part of the fried chicken family, a clan worthy of great respect in the Eatscape. Order wings with the sauce on the side, then sauce each wing just before you eat it. The wings will retain their crispy fried exteriors through consumption, and you'll get the exact amount of sauce you want on each one. Plus, if you're unsure which sauce to order, you can get two on the side, try each, then use the one you prefer.

circumstances, and in today's Eatscape, the ability to survive spiciness is not an adaptation that's going to get you very far. That's because the highly evolved among us understand the following:

1. A meal is meant to be *enjoyed*, not merely survived.

2. Spiciness should be a complementary characteristic, not a defining trait.

3. Eating is about pleasure, not feats of strength.

Of course, heat is relative and depends on the Eater, so I'm not saying there's an objective point at which foods become too spicy. Rather, I oppose the idea that heat is a desirable end in itself.

Consider Brendan's approach to Buffalo wings (the Forked Tongue), just one example of the many things people do to

TIP When eating wings, the best way to reduce napkin usage is not to use napkins at all. Get as messy as necessary, then go to the bathroom to wash up. To sip a beverage midway without dirtying your glass, see the Mid-Finger Pincer and the Palm methods in "Etiquette and Hospitality," pages 278–79.

offset the intensity of spicy foods. I don't understand why someone would go to such lengths to survive an eating experience they didn't have to undergo in the first place. If you enjoy heat at level seven, why start with a food that's level ten and then work so hard to reduce its effects? I'd rather have a degree of heat that offers sensation without suffering.

Over time, evolution will weed out those who place too much emphasis on spiciness by selecting for intelligence, not strength. So while Brendan is rubbing celery sticks together to put out the fire in his mouth, I'll be sitting in front of a glowing hearth, licking my fingers clean.

BIODIVERSITY AND HOMEOSTASIS IN A SNACK MIX ECOSYSTEM

Biodiversity is a sign of a thriving ecosystem that supports many different organisms, each making its own contribution to the Circle of Life. But if that diversity expands to include an especially fearsome predator, other organisms may be overpowered, upsetting the natural balance we call homeostasis. (Like the Ewoks say, "Biodiversity is great until Storm Troopers show up.")

These principles also apply to a good snack mix, one composed of diverse flavors that balance and play off each other. But if one component overpowers others in flavor, texture, size, or weight, homeostasis—and deliciousness—will suffer.

Balancing flavors and textures is largely a matter of taste, so I'll simply encourage you to combine contrasting elements without using any one snack that blocks out others. Maintaining homeostasis with regard to size and weight, however, is a more objective concern, because a good snack mix will have roughly the same distribution of all components throughout.

When small and large snacks are mixed or even shaken together, the small ones tend to fall to the bottom, leaving the large ones on top and producing a mix that isn't very well mixed. This phenomenon is known as granular convection, although it's sometimes called the Brazil Nut Effect. (Seriously. Google it.)

Bioengineering: Using the Brazil Nut Effect to Your Advantage with the Claw

The Brazil Nut Effect is a natural process—don't fight it. Use your understanding of nature's ways to bioengineer a well-mixed mix. Here's how:

Step 1: Prepare your mix in a clear glass bowl, so you can monitor all the strata as you work. Layer the biggest snack on the bottom and work your way up to the smallest.

Step 2: To mix the components, position your hand like one of those claws in the arcade game that you drop down into the bin of toys. (The one that never gets anything.) Reach down to the bottom of the bowl and bring your fingertips together, closing the claw while leaving space between all your fingers.

Step 3: As you lift your hand out of the mix, the Brazil Nut Effect will work its magic, and smaller pieces will move toward the bottom of the mix, lifting larger components toward daylight.

TIP Your snack mix should almost always include pretzels, which offer crunch, salt, and flavor without adding too much of any of them. The pretzel is the soil from which a bountiful and balanced mix may flower.

Repeat claw maneuver as necessary, making sure you mix gently and watch what you're doing. When it looks like you have homeostasis at all levels of your snack mix ecosystem, put away the claw.

GENETICS AND TAXONOMY

We've covered some visible examples of interdependence across the Eatscape ecosystem. But to truly understand the ways our foods are related, we must study their genetic makeup and categorize them based on their DNA.

DOMINANT AND RECESSIVE INGREDIENTS

Ingredients are the genes of the food world. So when breeding dishes, it's important to understand how to properly cross dominant and recessive traits.

Dominants are especially sharp, spicy, acidic, or bitter, and can overpower recessives. The sweetest or saltiest varietals, when used in excess, don't mask recessives as much as they just ruin them. Some of the most common dominant sandwich ingredients are mustard, pickles, raw onions, and roasted peppers. Mustard can be great in moderation. Pickles and raw onions are best when sliced very thin, not only because it reduces their dominance but also because it makes it easier to bite through them without pulling them out of the sandwich.

Roasted peppers are an interesting case. They're generally presented as a recessive complement to something else, but they dominate whatever they touch. So if you like them, that's fine, but don't squander expensive or precious ingredients in their company.

Indeed, that's a key lesson here. As much as many Eaters love meat and other delicacies, we don't love wasting them, and we don't love wasting money. Understanding how dominant ingredients work can help you reduce such waste.

For instance, if you smother a hot dog in mustard and sauerkraut, you don't actually like hot dogs. You like mustard and sauerkraut, with a salty, meat-like cylinder in the middle—which is fine. But in that case, you should really eat a veggie dog. With all those toppings, you won't know the difference, and you'll be saving a cow and/or pig, whose delicious hot dog parts deserve better.

The same goes for using sushi as a soy-sauce-and-wasabi delivery system. Eat an avocado roll. You'll save your money, and save the fish for the people who actually want to taste it.

THE SANDWICH GENOME PROJECT

Eaters at Sporkful University and across the world have recently embarked upon an effort to map the entire sandwich genome. It's not a small task. The work on genus *Grillescheesae* alone has already left more than one researcher lactose intolerant. The project will continue for years, but so far we've come up with one useful technique for testing sandwich gene pairs to see how they express themselves—the Punnett Square Sandwich.

PUNNETT SQUARE SANDWICH

If you want to better understand how sandwich genes pair together and you're not sure whether certain ingredients are dominant or recessive, try a combination. This way you can taste what happens when certain flavors are crossbred, without making whole sandwiches that you may end up disliking. If all quadrants taste good, you just created four sandwich strains worthy of reproduction.

Here's one example. Try your own test cross in your lab!

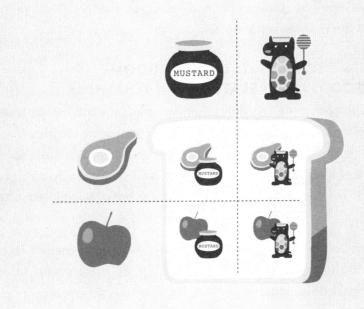

GASTRONOMY TAXONOMY: A GUIDE TO THE MENU OF LIFE

Now that we have a firm footing in genetics, let us turn to taxonomy, so we can use foods' genetic makeup to classify them properly. This effort is crucial because it helps you choose what foods to eat.

If you're trying to decide where to go for dinner, you might think, "Do I want Italian or French?" But in truth, this is a misleading way to think of your options. Some Italian foods are more closely related to some French foods than they are to each other, so a choice between nationalities is a false one. (Chicken parmigiana is more similar to chicken cordon bleu than it is to fettucine Alfredo.) Foods may share a cultural bond, but in terms of the eating experience, if you're craving chicken parm, pasta offers little comfort.

So before deciding between French and Italian, decide whether you want meat or pasta (or something else). If you want meat, perhaps then you'll decide between French and Italian options. If you want pasta, decide between Italian and Asian varieties. (If you choose Asian, you can then select from Chinese, Japanese, Thai, etc.) Good taxonomy clarifies your options and increases the chances you'll be pleased with your meal.

TO EACH ITS OWN KINGDOM: FOODS LESS RELATED THAN YOU THINK

As you no doubt recall, the field of taxonomy designates seven levels of classification for organisms, as follows: kingdom, phylum, class, order, family, genus, species. If you're like me, you memorized these terms in school with the mnemonic device "Ken poured coffee on Frank's gym shorts," which is a great way to learn taxonomy but a terrible waste of coffee.

Despite Eaters' best efforts, however, many foods are categorized incorrectly. Some are lumped together just because they share flavors, when they really should be considered distinct. Here are a few examples:

- **THIN-CRUST ("NEW YORK STYLE") PIZZA AND CHICAGO DEEP-DISH PIZZA** Eaters are always asking me to take a position on this long-standing food feud, but I refuse. Having lived in both cities and eaten my share of both styles, I've determined that they are different foods for different moods, each great in its own right.

- **TEXAS BRISKET AND JEWISH BRISKET** Just because you're starting off with the same cut of meat doesn't mean you're ending up with the same food. The former is smoked, the latter is braised, and the difference in the result is too great to be chalked up to recipe variation. There are some Texas Jews who claim to have crossbred these strains, but for the most part that appears to involve cooking a Texas brisket on Passover, which is more a sign of evolutionary progress than of artificial selection.

- **YEAST AND CAKE DONUTS** Yeast donuts are the softer, airier variety you may associate with Krispy Kreme. Cake donuts are dense and bready, like most of the options at Dunkin' Donuts. Lump these together and you might as well add in every other fried bread confection, from fritters to funnel cake.

- **CORN AND FLOUR TORTILLAS** This is the real dividing line when eating Mexican food, the first fork in the family tree. There's more of a fundamental difference between tacos made with corn tortillas and those made with flour than there is between quesadillas and burritos.

ALL IN THE FAMILY: FOODS SEPARATED AT BIRTH

Now let's study a couple of examples of the opposite phenomenon—foods considered to be unrelated but that are really kissing cousins.

- **ASIAN NOODLES AND ITALIAN PASTA** Long strands of starchy compounds boiled in water and combined with sauce, meat, and/or vegetables can be delicious whether their ancestors hail from the Orient or the Adriatic. As Marco Polo can tell you, there isn't such a difference between lo mein noodles and linguine by themselves—it's their preparations that separate them.

- **DUMPLINGS, PIEROGIES, GYOZA, EMPANADAS, ETC.** Whether you fill dough with meat and/or cheese and/or vegetables, and whether you boil, steam, or fry them, these are all wonderful expressions of a timeless concept.

- **QUESADILLAS AND BURRITOS (AS OPPOSED TO TACOS, TOSTADAS, AND TAQUITOS)** As previously stated, once you choose between flour and corn tortillas, everything that follows is a delicious variation on the same theme. So quesadillas and burritos are in one group here, tostadas and the like in another. It should be noted that enchiladas, by virtue of their emphasis on sauce and their utensil requirement, are just different enough to be considered an inbred first cousin.

Barbecue Potato Chips: A Lineage

I can't even begin to go through the taxonomy of all the many foods of the Eatscape, but perhaps by showing you the entire classification of one food—barbecue kettle-cooked potato chips—I can give you an inkling of this area of study:

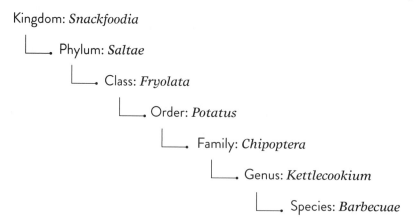

Kingdom: *Snackfoodia*
    Phylum: *Saltae*
        Class: *Fryolata*
            Order: *Potatus*
                Family: *Chipoptera*
                    Genus: *Kettlecookium*
                        Species: *Barbecuae*

PASTA PHYLUM CRITERIA: FORKABILITY, SAUCEABILITY, TOOTHSINKABILITY

The Starch kingdom is where you find phyla such as bread, rice, and of course, pasta. The Pasta phylum is broken down by shape into class Longus, class Shortus, and class Raviolus, much of which we'll cover soon. But first, let's look at the basic characteristics Eaters should consider when studying any pasta shape.

- **FORKABILITY** How easy is it to get the pasta on the fork in the desired amount and keep it there?

- **SAUCEABILITY** How well does the pasta take on and retain sauce? (See also "Sauce Adhesion: A Vital Consideration," page 262.)

- **TOOTHSINKABILITY** How easily can it be prepared al dente, which is Italian for "toothsinkable"? When cooked, does it tend toward flimsy or firm?

Examples of the best short pastas by these criteria include the grooved tube varieties, such as rigatoni, and the tight braid—gemelli. Cavatappi, a tubular corkscrew with ridges, may be the best of all. Workaday corkscrews like fusilli and rotini are adequate. They're very sauceable and forkable but turn soft quickly. Gimmicks like the wheel and the bow tie sit at the bottom of the short-pasta heap, as they tend to turn so mushy they're neither toothsinkable nor forkable.

In the long-pasta category, wide, flat pappardelle reigns supreme in toothsinkability and sauceability, but it can be so broad and thick that forkability becomes a challenge. Fettucine may provide the best happy medium. Angel hair sounds romantic, but its high SATVOR means it takes on liquid so quickly that it invariably turns to mush. Plus, it's a victory of marketing over logic. After all, would an angel's hair really be as thin as possible? When you see women in shampoo commercials showing off their flowing locks in slow motion, they always use words like "full-bodied" and "lush." Wouldn't an angel's hair be more like pappardelle, which is the epitome of that ideal?

PASTA CLASS LONGUS: CONSUMPTION STRATEGIES AND TOOLS

When composing a single forked bite of a long pasta like spaghetti, many factors are at play. Your goal is to combine all the desired parts of the dish—pasta, sauce, meat, veggies, etc.—into one bite, in your optimal ratio, in a structure and quantity that you can actually put into your mouth.

Here are some rules of thumb:

- **NO DANGLERS** Control quantity on the fork by sweeping horizontally across the pasta peaks, lifting to separate from the mass, then returning to the edge of the plate or a spoon to twirl (figure 9.2). Do not jab the fork straight down through the pasta to the plate and start twirling willy-nilly, unless you want a bite as sloppy as your technique.

Control Pasta Portioning with Horizontal Entry

When you stab straight down into a plate full of spaghetti, you have little control over how much ends up on your fork. You usually end up with too much to bite and danglers everywhere. Sweeping horizontally across the top allows you to choose exactly how many pasta strands to fork, and twirling those strands on the side of the plate ensures you don't pick up unwanted stragglers.

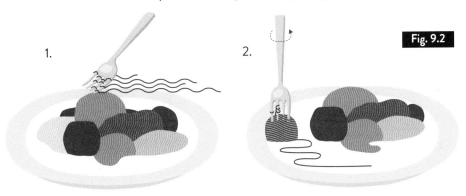

1.   2.   Fig. 9.2

- **A GOOD PASTA-TO-SAUCE RATIO** Once the pasta is secured on your fork, remove excess sauce by dabbing the bite on the side of your plate. Add more by sweeping the bite through the sauce.

- **MEAT AND VEGGIES INCLUDED AS DESIRED** You want to wrap pasta around the fork such that the tines remain exposed, so you can stab these items to add them to the bite. If you fail to do so, manually pull the pasta back to expose the tines (figure 9.3) or balance meat and veggies on top of the bite (figure 9.4). If you want more meat and veggies in the bite, stab those first, then use a utensil or finger to

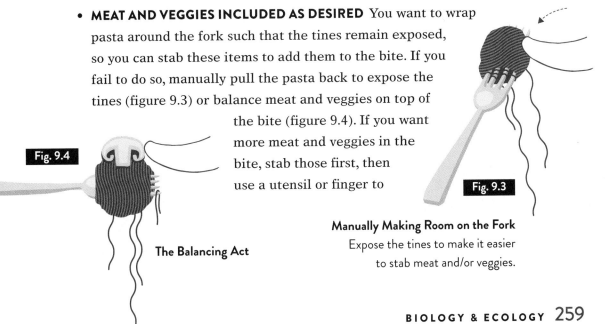

Fig. 9.4

The Balancing Act

Fig. 9.3

Manually Making Room on the Fork
Expose the tines to make it easier to stab meat and/or veggies.

slide them toward the handle to make room for pasta. (This is essentially the reverse of figure 9.3. You're moving the meat back to make room for pasta instead of vice versa.)

SHORT PASTA CONSUMPTION: DIFFERENT CHALLENGES, DIFFERENT SOLUTIONS

Short pastas are completely unlike long pastas. It's hard to believe they're so closely related, considering that transporting an ideal bite from plate to mouth requires such different techniques.

Slate's Julia Turner came on *The Sporkful* to argue for the superiority of short pastas:

> *I'm a short-shape partisan. They're easier to eat, hold sauce better, and are generally delicious. Once they're on the fork you can almost create a shovel [figure 9.5]. Scoop the short shapes around in the sauce, and get a little sauce on top of the pasta as well as intermingled with them and inside them. That's a delicious bite right there.*

Julia's right that the Shovel Formation is a good one, but I think she overestimates the ease with which short pastas can be firmly forked in the first place. When cooked al dente, hollow cylinder varieties have a tensile resistance that may cause them to eject from the fork after being stabbed. And the larger the diameter of the cavity, the less forkable they are. Sure, you can get one or two on the fork without much trouble, but if you want bites as toothsinkable as possible, you'll need more than that, which is where things get tricky.

As an alternative to the shovel approach you may try a tine insertion method called the Watson and Crick (figure 9.6), named after the two men who discovered DNA in 1953.

Fig. 9.5

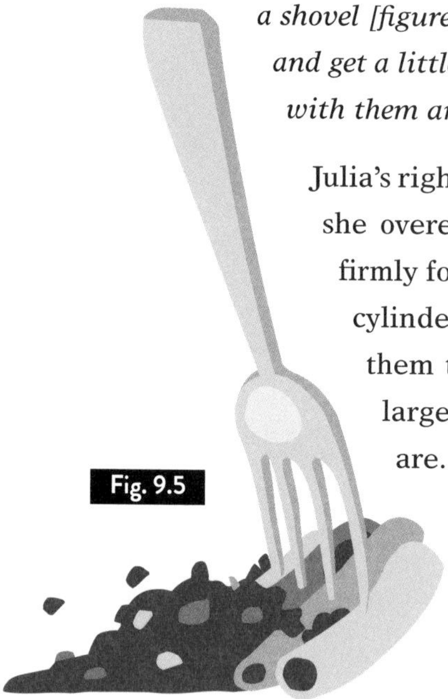

Short-Pasta Shovel Formation

The Watson and Crick Method of Short Pasta Consumption

This technique has the advantage of making it easy to get a few cylindrical pastas on your fork and keep them there. But you lose the scoopability of the Shovel Formation, and if the pasta tends toward the longer end of the short spectrum, you've covered the tines and rendered them useless for stabbing meat and veggies.

All things considered, I believe the Shovel Formation is better than the Watson and Crick, but it puts great pressure on the cook. If short pastas are a tad too firm, forkability suffers. If they're a tad too soft, toothsinkability declines.

Fig. 9.6

WHY CLASS LASAGNAE ISN'T WORTH IT

There are almost infinite ways to bring together pasta, sauce, cheese, and meat, and lasagna is the most painstaking and tedious of them all. To make it worth the effort required to separate those cooked sheets of pasta and layer the components just so, lasagna would have to offer the Eater some extraordinary attribute not available in any similar constitution of the same ingredients. It does not.

Now if someone else wants to make lasagna for you, and their time would not otherwise be spent bettering your life or the larger Eatscape, there is no reason to stop them. It's true that lasagna is delicious and beautiful, and destroying something that's been built with such care has its joys.

But if you're making it yourself, baked ziti is a superior alternative. It allows you to combine the same ingredients in the same proportions, except in a manner that's far less time consuming to assemble, if also less attractive. So instead of lasagna, make baked ziti, or better yet, baked rigatoni, which has superior sauceability. (See also "Sauce Adhesion: A Vital Consideration.")

SAUCE ADHESION: A VITAL CONSIDERATION (OR, WHY ZITI IS NOT THE BEST PASTA FOR BAKED ZITI)

Hey, look, it's our old friend surface-area-to-volume ratio! Ziti and rigatoni are almost identical pasta shapes—same genus, different species. Both are hollow cylinders, roughly 1.5 inches long, and about a quarter to a half inch in diameter. But there's one crucial difference: Rigatoni has exterior grooves.

Those striations increase rigatoni's SATVOR and boost its adhesive power. For this reason you should really never use a smooth-sided short pasta when one with grooves is available, as sauces of all kinds will cling more readily to the ridged varieties. Baked ziti is no exception.

RANKING CHICKEN PARMIGIANA SERVING METHODS

Chicken parm has little to do with pasta, even though they're both found at Italian restaurants. I'm including my discussion of chicken parm here first to reiterate that point, and second, in case all this talk of starch has you craving fried meat.

The members of the genus *Chickenparmae* are closely related but far from identical. Which one is best? The primary concerns are to preserve as much chicken crisp as possible, and to attain proper ratios between the fried chicken breast, cheese, and sauce. That means all three coming together into a greater whole, an ensemble cast where no one star emerges.

With those criteria in mind, here's how the options stack up, from best to worst:

1. **SUB** Keeps chicken edges exposed to air to maintain crisp. Easy to invert to bring cheese closer to tongue. Order it with the sauce on the side and dip on a per-bite basis to preserve crisp and manage ratios.

2. **CALZONE** I'm talking about fried chicken breast, sauce, and cheese inside dough. It's more common in some regions than others. It loses all fried crisp but replaces it with exterior calzone crisp, and the interior is moist and cheesy.

3. **PIZZA** Great in theory because the chicken looks so crispy, but the meat is usually fully cooked before being placed on the pizza. When the pizza is baked the chicken dries out.

4. **ENTRÉE** This option with a side of pasta may be the classic, but it's the worst. The chicken ends up sitting in sauce and loses all crisp. Plus the meat tends to be cut thicker, thus throwing off ratios to sauce, cheese, and fried.

BIOENGINEERING LAB

As I've said, you have a lab in your home—you just call it the kitchen. This section includes some of the more experimental techniques and dishes in the Eatscape today. I encourage you to tinker with them in your own labs and share your results.

THE POPCORN MOMENT

You may have heard of the eureka effect, the experience of having a sudden realization or discovery. In the Eatscape we call it a Popcorn Moment.

The next time you're lacking in Eater inspiration, think about popcorn. Imagine the first time, thousands of years ago, when someone—presumably by accident—dropped some dry corn onto a fire. Not only did it start exploding, but the shrapnel was delicious.

Popcorn is so simple yet so radical. It epitomizes the spirit of experimentation and *joie de eating* that we prize here at Sporkful University. Here are some discoveries I've made during my own Popcorn Moments.

In Vitro Popcorn-Butter Fertilization

When you have butter pumped on your popcorn at the movie theater, it's rarely spread evenly throughout. More thoughtful theater employees will pump some butter when the bucket is half full, then more when it's filled, but the result remains imperfect.

Sometimes a theater leaves the butter station on the customer side of the counter and lets you pump it yourself. If that happens to you, use soda straws to jury-rig this in vitro popcorn-butter fertilization kit to achieve consistent buttering. Reposition the bag after each pump so you can fertilize all areas of popcorn equally. If you have an especially big bag of popcorn, you may need to double up on straws to reach the bottom. And theater owners, take note—this should be the norm.

Popcorn Consumption Techniques

The Bullfrog

This is my personal favorite way to eat popcorn. You will never feel more power over your food than when you eat it this way. I wish I were a frog.

The Face Funnel

This is a more gluttonous approach that shows that popcorn isn't just fun to make, it's also fun to eat.

RANKING POPCORN PREPARATION METHODS

1. **LARGE MOVIE THEATER OR CARNIVAL-STYLE POPPER** Assuming it's fresh and not too salty, this is the best. It's never burnt, it's still warm, and it's well imbued with seasonings.

2. **COOKED IN OIL ON THE STOVE TOP** This is the best way to assimilate seasonings into the popcorn at home.

3. **AIR POPPED** It has a light, crispy texture, but the butter (or oil) and seasoning never quite adhere the way you'd like. Either it's too dry or too greasy.

4. **MICROWAVE** The buttery substitute and seasoning are soaked in pretty well, but you always end up with too many kernels burnt and/or unpopped, and the texture is a little off. And because popcorn lung.

SPORKFUL HALL OF ELDERS INDUCTEE: EDDIE GRAIS

Every university needs a way to honor the members of its community who have made the greatest contributions, those who have provided the most awe-inspiring Popcorn Moments. At Sporkful University, induction into the Hall of Elders serves as that honor.

Eddie Grais (pronounced "grace") "was a real party boy, larger than life," says his daughter, Karen Grais Meyer.

Karen says her father always loved salami. In the early years of his marriage to Karen's mother, they'd drive from Chicago to Michigan City, Indiana, with friends, and pass salami sandwiches to each other out the windows of their cars. When he needed a parking ticket fixed, he sent the cops in Chicago a salami.

It makes sense, then, that Eddie Grais's famous salami creation would be a part of his legacy. The details are hazy, but at some point in the sixties, at one of the many barbecues Eddie loved to host, he introduced the dish that has earned him induction into the Hall of Elders.

In that moment, Drunken Salami was born.

Drunken Salami is marinated for at least three weeks—often much longer—in equal parts Scotch and Russian dressing, then grilled to boozy perfection. Some of the alcohol cooks off as the dressing caramelizes and the outside chars.

The recipe has only been shared with a select few over the years—I learned of it from Karen's friends Michael and Carla Levin, and their daughter, Rachel. But it has never been made publicly available . . . until now.

As for Eddie, he never lost his passion for people, or food. When he turned seventy, he knew he was sick and might not live too long. He threw himself three birthday parties. Eddie passed away soon thereafter, in 1990.

"He had a bad heart and I'd go to pick him up at the hospital and he'd say, 'Let's go out to lunch,'" recalls Karen. "He had a button he wore the last year of his life that said, 'Don't postpone joy.' That was a real motto of his."

I think it should be a motto of ours as well. So it's with great pleasure that I hereby induct Eddie Grais into the Sporkful Hall of Elders. And with Karen's permission, I am pleased to share the recipe for Drunken Salami, with a couple of my own tweaks, for all the Eatscape to enjoy.

DRUNKEN SALAMI

YOU WILL NEED

2 kosher beef salamis (bullet shaped—about 6 inches long and 2.5 inches thick)

1 cup Russian dressing

1 cup Scotch

INSTRUCTIONS

Mix Scotch and Russian dressing. Remove all wrapping from salamis, but do not otherwise alter or pierce them. Place salamis in a plastic bag, pour in the liquid, and store in the fridge. Rotate salamis and mix marinade every week for at least 3 weeks, and up to 4 months. Karen says, "The longer you marinate it, the better it is."

When you're ready to cook, set up your grill to provide both direct and indirect heat. For charcoal grills, bank the coals steeply to one side, about 3 inches below the grates at their highest point, and open top and bottom vents halfway. For gas grills, set the primary burner to high and turn off all other burners.

Remove salamis and wipe off most of the excess marinade with a paper towel. Place them on the cool side of the grill and cook until the interiors register 125° to 135° degrees (20 to 25 minutes), flipping and rotating salamis halfway through cooking. Then slide them to the hot side of the grill (if using charcoal, open bottom vent), and cook, uncovered, until well browned and charred on all sides, 6 to 8 minutes.

Once you remove the salamis, slice them in half the long way, then chop into half-inch thick semicircles and serve. "You should taste the Scotch a little bit," Karen explains.

NOTES

- When in doubt, err on the side of cooking it slower and lower.

- Russian dressing means different things in different parts of the country. In this case it's referring to a red, syrupy dressing similar to Catalina, not the mayonnaise-based dressing similar to Thousand Island.

- If you don't want to have to rotate the salami regularly in the fridge, you can use twice as much Scotch and dressing, so the salamis are more fully submerged. (That does make the dish more expensive.)

- Drunken Salami is not traditionally eaten with any condiment, and although I don't think it needs one, I've served it with yellow mustard or honey mustard to positive reviews from Mrs. Sporkful.

THE TRAGIC PIÑA COLADA RECONSTITUTED

Here's another example of a great advance from a passionate Eater working in a lab not unlike yours.

When MSNBC's Rachel Maddow joined me on *The Sporkful*, she really wanted to talk about the tragedy of the piña colada. It's a story rooted in drinking history, and as Rachel pointed out, "drinking history was written down by drunk people. So it's all a little fuzzy." (Not unlike salami history.)

The classic piña colada was supposedly invented in Puerto Rico in 1957, eight years after the advent of Coco Lopez, the cream of coconut substance that serves as the base for the piña colada. Coco Lopez is pretty gross—a "chemical slurry," as Rachel puts it, full of artificial ingredients with really long names. She explains why the piña colada is different from, say, a daiquiri:

> When we think about bad chemical drinks that give you a stomach-ache, make you fat, rob you of your brain cells, and make you feel like you've eaten Twinkies, we think about having redesigned drinks that were once wholesome.
>
> The tragedy of the piña colada is that it was never wholesome.
>
> This makes it unlike other drinks that have been made into awful things in modern times, like the daiquiri. Now you're likely to get it out of a Slurpee machine, and it's made with some chemical stew that makes Coco Lopez sound like mother's milk.
>
> But you can reverse-engineer a daiquiri into what it was originally, which was rum, lime, sugar, end of story. And it's a wonderful drink.
>
> There's no way to do that with a piña colada.

Until now.

THE MADDOW COLADA

With Rachel's blessing I am happy to share the recipe for the Maddow Colada, printed here alongside the traditional piña colada recipe for comparison purposes only. If the piña colada had originated as a wholesome and delicious beverage, this would have been the recipe.

ORIGINAL PIÑA COLADA

1½ ounces light rum

1 ounce Myers's dark rum

2 ounces Coco Lopez

1 ounce cream

4 ounces fresh pineapple

THE MADDOW COLADA

1½ ounces light rum

1 ounce 8-year-old medium-bodied amber rum

1½ ounces unsweetened coconut milk

1½ ounces orgeat (almond syrup)

4 ounces pineapple juice

Fresh pineapple wedge, for garnish

INSTRUCTIONS

Blend rums, coconut milk, orgeat, and pineapple juice with ice. Don't strain. Serve in a short glass garnished with pineapple wedge and tiny umbrella. Serve with short, stubby (fat) straw.

Rachel correctly points out that coconut milk doesn't have the same rich, sweet mouthfeel you get from cream of coconut. That's why the almond syrup is the key. (Organic brands are available online. She recommends Teisseire.)

SCIENTIFIC STUDIES FOR WHICH SPORKFUL UNIVERSITY IS REQUESTING FUNDING

If you read the news, you know there seems to be no shortage of money available for research designed to tell us things we already know. In recent years I've seen reports confirming that cheese tastes better melted, beer makes your brain happy, and bacon sandwiches help cure hangovers. (I'm not sure which poor souls were forced to consume enough melted cheese, beer, and bacon to make these studies statistically valid, but we owe them a debt of gratitude.)

I've looked at the grant proposals that got funding for those studies and determined that to gain access to coveted research dollars, we must engage in the time-honored academic tradition of making obvious and simple concepts sound revolutionary and complex. It's actually pretty easy. All you need are some ideas and a thesaurus.

I'll soon be requesting funding for the following studies:

- I want to examine whether food tastes better when it's free. This means I'll need to use grant money to pay for half my meals, so I can compare them to the meals I purchase myself. This study will be called "The Relative Sapidity of Unrecompensed Comestibles."

- I have a sneaking suspicion that very strong coffee increases my energy level and mental acuity. When consumed in large quantities, it may also grant me superpowers. I'd like to simultaneously confirm this hypothesis and defray the cost of my habit with a study entitled "The Cerebral Concomitants of Brobdingnagian Caffeine Quaffing."

- When I had summer jobs, I learned that beer tastes better after manual labor. I'd like to find out whether that's objectively true. And because I'm so devoted to the cause of science, I'll even agree to be in the control group, which means I'll drink free beer *without* doing manual labor. I'll call this study "Corporeal Exertion and Pro Rata Pint Delectation."

Of course, you have a better shot at funding if you have a track record of success. So let's look at one more discovery from the SU lab.

THE TOT TO TROT

This sandwich is my Frankenstein. It began as an attempt to create the most complex egg and cheese sandwich imaginable, and to give it a terrible pun for a name. It ended with a resounding success on both counts. It's a grilled ham, cheese, and tater tot sandwich encased in an egg patty and placed inside a grilled cheese sandwich.

Make a grilled ham, cheese, and tater tot sandwich with the tots extra crispy and set aside. Scramble at least 10 eggs and add salt and pepper to taste. Heat and butter an 8-inch nonstick pan and fill pan with a half-inch-thick layer of eggs. Once eggs have started to solidify, flip them and place the ham, cheese, and tater tot sandwich on eggs. Ladle more eggs into pan, all around sandwich. Tilt pan in all directions and let sides of pan cook eggs so eggs form to the mold of the pan and solidify around sides of the sandwich.

Pan tilting and eggs cooking
You're using the sides of the pan to both cook the eggs and mold them around the sandwich.

When eggs are solidified, remove structure from pan and place on a plate, flipping it as you remove it so the egg bottom is facing up on the plate. Pour remaining scrambled eggs into pan. Let them solidify a bit and flip them, then return the previous structure to the pan, egg side up, the same way it was sitting on the plate. Tilt pan in all directions again so that raw eggs cook and fuse to cooked eggs on the sides of the sandwich. Remove and set aside.

Cook an open-faced grilled cheese sandwich with full cheese coverage on inside of both slices of bread. Place the egg-encased structure inside and close the grilled cheese sandwich. Slice it open and marvel at the cross section. It should look something like this:

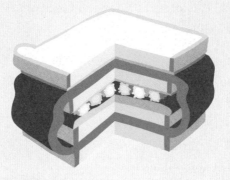

HOMEWORK

U nderstanding the large and small ways in which we and our foods are interconnected can lead to a more delicious life, as well as innovations and discoveries that further enrich us all.

Your assignment for this chapter is simple to explain yet difficult to execute.

Channel Charles Darwin, Gregor Mendel, and Elder Eddie Grais to create a truly new dish, or improve an existing dish substantially enough that it warrants a new name. In your kitchen/lab, run test crosses and ensure that the fittest results survive. Take risks. Experiment until you get it right. If you don't have at least one failure along the way, you're not trying hard enough.

Report your results to me at dan@sporkful.com.

10

ETIQUETTE & HOSPITALITY
When Eaters Mingle

- Hygiene, napkins, and bib use

- Napkin and bib use

- Scented soap and food pairings with a soapellier

- The Moist Towelette Museum

- The best types of gum and gum packaging

- The Eater-eatery relationship

- Hosting events large and small

- The Tippling Point

- Unorthodox wedding planning and the problem with wedding cake

- The Cocktail Hour Buddy System

- Buzz management

You've learned how to eat more better, but how do you put your new-found skills to use when interacting in the larger Eatscape? In her book *Emily Post's Etiquette*, Emily Post defines the "Best Society" as an "unlimited brotherhood" we should all strive to join:

> *Best Society is not a fellowship of the wealthy, nor does it seek to exclude those who are not of exalted birth; but it is an association of gentle-folk, of which good form in speech, charm of manner, knowledge of the social amenities, and instinctive consideration for the feelings of others, are the credentials by which society the world over recognizes its chosen members.*

Sound familiar? She might as well be describing the Eatscape. We've talked about the importance of speech in "Language Arts" and consideration for others in "Biology and Ecology," among other chapters. And throughout we've covered the most important social amenity of all—great food.

It's time to cover charm of manner. If the Eatscape is our own Best Society, then the truly refined among us are those who pursue Perfect Deliciousness most passionately, and facilitate others' pursuit of it. Conversely, the crudest behavior of all is to thumb your nose at Perfect Deliciousness while in the company of other Eaters.

That means a restaurateur must know how to use scented soaps correctly or risk poisoning our palates. A host must appreciate the importance of blood flow and course pacing or risk crushing a party's energy under its own excess. And a guest must employ good buzz management or risk ending up too full and/or drunk, or not full and/or drunk enough—both signs one has lost sight of Perfect Deliciousness and thus grave insults to the host.

With repetition, the rules of etiquette become second nature. As Emily Post puts it, good manners "must be so thoroughly absorbed as to make their observance a matter of instinct rather than of conscious obedience."

When that happens, we'll all taste the difference.

HYGIENE

Eaters don't mind getting messy when it's a necessary by-product of deliciousness, but mess in itself is not a desirable end. If you can eat just as well and keep yourself clean, that route is preferable.

Poor table manners may cause foods to mix improperly, through sloppy hand and face contact and substandard napkin technique. Ineffective soap methodology will further muddy your mouth's waters. And this kind of behavior has a negative impact on others. After all, watching a fellow Eater act so uncivilized will turn your stomach.

THE PERILS OF SCENTED SOAPS

The idea of soaps and other cleaners made from natural ingredients has an obvious appeal to the anti-carcinogen crowd, but its dangers are less apparent. When I go to the store, I can't tell the difference between the grocery section and the shampoo aisle. Pomegranate Creamsicle? Peanut Butter Cookie? I go shopping for hand soap and end up starving.

The real problem with any scented soap is that if you use it right before you eat, its smell remains on your hands and affects the taste of the food.

That's why thoughtful restaurants will either use unscented soaps or offer a selection of soaps to be paired with specific dishes. I recommend the appointment of a soapellier to oversee this portion of the menu. (See figure 10.1 for proper soap service and "The Soapellier Suggests" for pairings.)

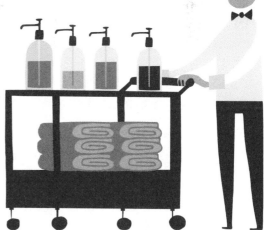

Fig. 10.1 Proper Soap Service with a Soapellier

THE SOAPELLIER SUGGESTS: SOAP AND FOOD PAIRINGS WE CAN SUPPORT

If restaurants insist upon using scented soaps, they should put more thought into how those scents pair with their foods. Hiring a soapellier can help. Here are some possible pairings:

- Honey soap with ricotta cheese
- Melon soap with prosciutto
- Lavender soap with ratatouille
- Lemon verbena soap with diet cola

NAPKIN TECHNIQUE: CLOTH VS. PAPER, QUADRANT SEPARATION, AND SEAM USAGE

Cloth napkins are far more durable than paper ones, which are insufficient lap shields in the presence of sauce. Cloth is also more environmentally friendly. But overstarched cloth napkins, and those made of synthetic fibers, don't absorb drips and wipes very well. Plus when the fabric is dark in color, maintaining quadrant separation becomes difficult. And they don't feel like something you're supposed to be wiping your dirty hands on. You might as well be smearing sauce on someone's drapes.

Paper napkins, on the other hand, have a rougher texture that removes superficial mess more effectively—in part because you can always get another one—while also *feeling* more like that is their express purpose.

Whatever you're using, use these strategies:

- **QUADRANT SEPARATION** Divide your napkin into four equal regions and use each separately. This extends the life of your napkin so that even when you're three-quarters of the way through a meal, you still have a bit of clean surface available.

It also guards against inadvertent cross-contamination, because it offers the ability to use different areas of the napkin for different food messes.

- **THE SEAM** There is one area of the cloth napkin that is rougher than any paper napkin, and thus best suited to removing tough grime: the seam. When tackling tough stains, clean the affected area as much as possible with the napkin proper, then use the seam to remove the most stubborn hand and face blemishes. (Exercise caution, however, because food residue left so close to the napkin's edge is likely to end up on your clothes.)

- **BIB UP** Bibs should be considered acceptable for all people all the time—not just children and lobstervores. A true bib is tied around your neck and used in concert with a lap napkin. A napkin tucked into your collar is a poor substitute, as it leaves your lap unguarded, and its ruffles interfere with arm movement.

MOIST TOWELETTES: USE AND DISPLAY

Moist towelettes are far from perfect. Many don't smell great, the cheap ones fall apart easily, and they're no friend to sewer systems. But some messes just won't unmess without moisture, and when you're on the go, they're often your best option.

To learn more about these sanitation devices, I talked to John French. He's the interim director of the Abrams Planetarium at Michigan State University, but more impressively, he's the founder of the Moist Towelette Museum.

The MTM is housed on the Michigan State campus, in John's office, where it occupies about eight shelves. Each towelette is displayed on a black cloth. It's a rotating collection, of course, because John doesn't have space to display his thousands of towelettes. Admission is free, and John says the museum welcomes anywhere from one to two visitors each year. (You can get a taste of the place on its website, where moist towelette philanthropists may also donate to the collection.)

An entire wing (read: shelf) of the museum is dedicated to restaurant towelettes, another to airline towelettes. But the odd ones are John's favorites. He has a moist towelette that says it's for after mammograms, one that promises temporary relief of hemorrhoids, and another that claims to be for wiping away radioactive contamination.

Clearly, this guy knows a thing or two about moist towelettes. Here are his tips for selecting and using them well:

- Cloth-textured ones are more durable than papery ones.

- Lemon-scented is better than chemical-scented.

- Alcohol-based is better than soap-based. Your hands dry up faster, without a soapy residue, and even the towelette itself will dry quickly, providing you with a napkin for additional cleaning.

I would add, only use moist towelettes after you're done eating, because, as with hand soap, you don't want their scent to affect the taste of your food. Most importantly, follow the instructions carefully: "Tear open packet, remove towelette, and use."

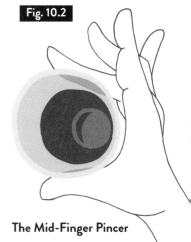

Fig. 10.2

The Mid-Finger Pincer

SANITATION WHEN MOVING BETWEEN FOOD AND DRINK

The sight (and taste) of a half-full glass clouded by greasy fingerprints is demoralizing enough to make the glass seem half-empty. Fortunately, you can prevent such horrors.

If you're elbow-deep in ribs or wings and want a sip of your beverage, no napkin technique will save you, which is where these two maneuvers come in.

The Mid-Finger Pincer (figure 10.2) is preferable, because it's less clumsy. But if your fingers are fully coated, The Palm (figure 10.3) is your only option.

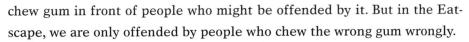

Fig. 10.3

The Palm

BREATH MAINTENANCE: ON CHEWING-GUM SHAPES AND PACKAGING

In the etiquette community, gum is a controversial issue. Miss Manners says you should never chew gum in front of people who might be offended by it. But in the Eatscape, we are only offended by people who chew the wrong gum wrongly.

What shape, eating method, and packaging are best? *Slate*'s Mike Pesca is an avid gum chewer. When he came on *The Sporkful*, here's what he said:

> *I think that if I were a very rich man and could have bowls full of gum, I might choose the Chiclet shape [the small, pillow-shaped rectangle with a hard exterior]. But I just think the Chiclet shape is another version of life's "that's how they get you." You're paying more for less. I need two Chiclets as a gum experience, but one stick does the job for me.*
>
> *So in terms of economy and fun, I enjoy the stick, because you get the double-fold action.*

That last point is a big one. Everyone feels like a gorgeous, scantily clad supermodel when you fold a stick of gum over on your tongue seductively, just like the ladies in the commercials. If you're feeling bad about your appearance, a little double-fold gum action should help you get your groove back.

But chewing isn't the only concern. Portability is also important, since you often carry gum with you. The increasingly popular blister pack is portable and keeps gum from falling out, but it's a large container without much inside.

Pesca argues that the most problematic trend in gum packaging has been the introduction of the valise container. It seems elegant upon purchase, but its latch quickly disintegrates, at which point there's no countervailing force to hold the remaining gum in place (figure 10.4).

Problems with the Gum Valise—and a Solution

This variety of gum packaging looks nice when you buy it, but it doesn't hold up over time.

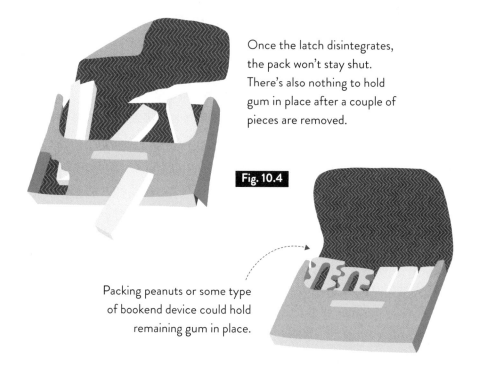

Once the latch disintegrates, the pack won't stay shut. There's also nothing to hold gum in place after a couple of pieces are removed.

Fig. 10.4

Packing peanuts or some type of bookend device could hold remaining gum in place.

The problems of the valise highlight the genius of more traditional gum packaging. When it opens from the side, as in a five-pack of Bubblicious, pieces are held in the wrapping until you peel it back (figure 10.5). When it opens from the top, as in the Plen-T-Pak, paper stays behind to act as packing material, so remaining pieces are held in place (figure 10.6).

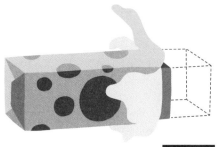

Fig. 10.5

Classic Peel-Back Wrapping
This packaging keeps a tight grip on
the gum until you're ready to chew it.

Fig. 10.6

**Plen-T-Pak with Built-In
Packing Materials**

Whatever type of gum you choose, it's important to show fellow Eaters that you've put thought into the decision and chosen wisely for your purposes. As long as you do, you won't be offending anyone.

ETIQUETTE IN THE
EATER-EATERY RELATIONSHIP

They say the customer's always right, but we all know that's not actually true. If you walk into a restaurant and demand to go into the kitchen and cook your own meal, you're wrong. So where do we draw the line? What should you expect from restaurants and bars, and what should they expect from you?

MENU MODIFICATIONS:
HOW FAR IS TOO FAR? HOW FEW IS TOO FEW?

An Eater in New Orleans who goes by the name Kicker called in to *The Sporkful* seeking guidance. When he goes out to eat, he has so many special requests that he's often asking the restaurant to create entirely new menu items just for him, and others at his table find it embarrassing.

As I told Kicker, extreme menu meddlers should abide by the following principles:

- When making special requests, always ask politely and be willing to pay extra.

- It's always acceptable to modify, swap out, add, or subtract one ingredient in any dish that's on the menu.

- While restaurants should do their best to accommodate any request, you never have the right to *expect* them to make you something that's not at all on the menu.

On the other side of this discussion are those restaurants that forbid modifications or substitutions of any kind, no matter how minor. It's almost always because their priority is on turning out food as quickly as possible, and special requests slow down the kitchen and require a higher-caliber staff.

One restaurant in New York informs patrons, "Our sandwiches were created to be enjoyed as they are. Please, no modifications!" The wording suggests a chef's intent, even though their menu suggests it's more about efficiency. Every one of their fifteen sandwiches has the same condiment and veggies. What's so creative about that?

If an establishment believes it must ban menu modifications so it can serve customers faster, it has that right, just as people have the right to go elsewhere. But it's improper for eateries to suggest that the policy is rooted in some grand culinary vision when in fact it's rooted in expedience.

THE EATABILITY REQUIREMENT: TALL SANDWICHES, SHELLFISH, AND SLOPPY FOODS

Most restaurants serve food that's edible. But is it *eatable*?

Eater Wayne Hammond of Ashland, Oregon, defines eatability as "the compatibility of the food with the Eater's interface." He called in to *The Sporkful* to rail against absurdly tall sandwiches and saucy shellfish dishes served with the shells still on. Eaters Jonathan Goldberg and Kellie Fitzgerald of New York called in to debate what responsibility a restaurant that serves messy food has to help you deal with that mess.

There's no grand theory of eatability, as it really depends on the particular dish. But here's my take on the aforementioned situations:

- **ABSURDLY TALL SANDWICHES** If the fillings are all one ingredient, like pulled pork or deli meat, there's no such thing as too much, because you can always take some out and eat it separately without compromising the essence of the dish. But if a sandwich has so many different ingredients that you can't physically get them all in one bite, then the sandwich, as its creator envisions it, is uneatable.

- **SHELLS ON SHELLFISH** Separating shellfish from their shells is laborious and, when sauce is present, very messy. It's true that the shells add flavor to a sauce, but there's no reason why the restaurant can't remove them after cooking and before serving. (The minority who likes to eat the shells could request "shells on.") In a multifaceted dish, clam and mussel shells do provide the theoretical advantage of making the shellfish easier to find, but in truth, many of them become dislodged anyway. Then the empty shells torment the Eater with false hope, like working lighthouses on deserted islands.

- **SLOPPY FOODS** Most of the time you can tell when a dish will be sloppy, so it's up to you to take adequate precautions. (If a restaurant serves a messy version of a traditionally neat food, that's another story.) You should make sure you have napkins and a place to sit and eat. But if you're dining in, it's up to the eatery to serve the food on a large enough plate to contain drips and drops.

> **TIP** Serving clams in the shell may help show they're safe to eat, but if you're really in doubt, you probably shouldn't order the clams anyway.

BALANCING THE IMBALANCE BETWEEN
BARTENDER AND BARTENDEE

There's an understanding in the restaurant world that bartenders rule the roost. To some extent this is justified, because bartending requires many talents.

Chefs have to cook, but they don't have to deal with customers. Servers need people skills and menu knowledge, but they don't actually prepare much food. Bartenders must possess all the expertise of a server plus the ability to make great drinks, all while standing sentry over the business's profit center—the alcohol.

Top-notch mixologists deserve their exalted status. The problem is, there are far more bars than great bartenders. Too many of these people seem unaware that, no matter how good they are at their jobs, how trendy their watering holes, or how badly we want the sweet, sweet nectar they possess, bartending remains a sector of the *hospitality* industry.

Substandard bartenders violate the rules of hospitality in one or more of the following ways:

- **HOLDING THE BOOZE FOR RANSOM** You should receive good service and tip well when you get it. You should not have to tip well first in order to receive good service afterward.

- **GETTING LOST IN CONVERSATION** Bartenders chatting with patrons is a time-honored tradition, but it must be balanced with other duties. You want to finish a thought? No problem. You want to finish this epic tale about the time you skied down Mount Everest while tripping on acid? Pause it. I'm thirsty.

- **LOSING TRACK OF THE LINE** When a bar is packed three deep, good bartenders will start at one end and work their way to the other, then jump back to the beginning, so that people are served in order. A bad one will skip around, favoring the most aggressive and/or attractive in the crowd.

- **IGNORING YOU** Having waited tables for many years, I can tell you that intentionally failing to see customers is a real craft, but one that should only be practiced when absolutely necessary. If you feel like you're being ignored on purpose, you probably are.

A truly great bartender is a special person, worthy of adoration. But the best in the world still needs patrons.

EVENT PLANNING AND THE IDEAL HOST

There are two types of hosts: martyrs and missionaries. Martyrs sacrifice their own enjoyment so that their guests may be happy, deriving their pleasure from the knowledge that others are having fun. Missionaries lead by example, setting a tone of revelry and merriment that becomes contagious.

Of course, all good hosts look after guests' needs. But in the ideal scenario you won't just *oversee* your event, you'll also *attend* it.

> TIP Want to send an early signal to guests that you're on a mission to make your event a success? Answer the door with a drink in your hand.

BLOOD FLOW AND PACING: THE KEYS TO PARTY SUCCESS

When planning any type of function, the most important issues to consider are guest blood flow and food pacing. When people consume large quantities of food and drink while sitting, they become full and tired, and the collective energy of the gathering sags. When you space out courses and keep people moving around, it aids digestion and blood flow, boosting the energy level of your guests.

This is especially important at events with lots of courses and huge quantities of food. Many weddings today feature cocktail hours, which, though delicious, constitute whole meals in themselves. Then guests are ushered into a main hall, where salads are waiting at each place setting.

I just inhaled three dozen pigs in blankets. Now you want me to sit down and eat a pile of lettuce?

The salad course at large functions should be eliminated entirely. It's usu-

TIP A buffet dinner is an underrated option at large functions. It lets guests decide what (and how much) they eat, and provides another opportunity to get people out of their chairs and moving around.

ally uninspired, and worse, it forces guests to sit down and eat more at the crucial moment when they're feeling a little tipsy and a little full, and party momentum is balanced on a butter knife's edge.

After the hors d'oeuvres, there should be at least one hour during which drinks are available but no food is served, when guests are encouraged—implicitly or explicitly—to stand up. This may mean music and dancing, mingling by the bar, or any other activity that promotes blood flow and digestion. After this food moratorium, dinner may be served.

THE BENEFITS OF THE PRE-CEREMONY COCKTAIL HOUR (OR, THE MARRIAGE OF MRS. SPORKFUL AND ME)

Since you asked, I'll tell you about my wedding to Mrs. Sporkful. We did our cocktail hour and hors d'oeuvres before the ceremony, which offers a few benefits:

- It loosens up the crowd and lets guests who may not know many people make friends.

- Moderate quantities of passed hors d'oeuvres at the outset take the edge off guests' appetites and lay a foundation for forthcoming alcohol.

- After the ceremony there was another hour or so of mingling, drinking, and dancing, so that by the time dinner was served, people were ready to sit down and eat, and hungry enough to actually enjoy the buffet of fried chicken and mac and cheese.

STANDING UP AGAINST THE SIT-DOWN DINNER

When you get married, your son has a Bar Mitzvah, or your daughter has her debutante ball, everyone is going to tell you there are all these things you *have* to do.

"You *must* have cake!"

"You *must* serve a sit-down dinner!"

"You *must* dance the [insert traditional dance of your ethnic group here]!"

In truth, at your wedding, the only thing you *must* do is get married. At your Bar Mitzvah, the only thing you *must* do is become a man. And at your debutante ball, the only thing you *must* do is introduce yourself to society as an attractive, submissive young lady ready to occupy the traditional gender role assigned to you by your paternalistic forebears.

Everything else is optional.

If you want to have maximum blood flow and food pacing, eliminate the sit-down dinner entirely. Make small but substantial foods (like sliders) available throughout the duration of the party, which allows people to pace themselves and mingle and dance without interruption. (It's also usually less expensive.)

ELIMINATING WEDDING CAKE (AND OTHER TIPS ON IMPROVING DESSERT)

Mrs. Sporkful and I bucked convention at our wedding in more ways than one. We didn't serve wedding cake. Here's why:

- Most wedding cakes just aren't that good. Don't even get me started on fondant. (See "Fondant? More like Fondon't!," in which I get started on fondant.)

- Cutting it up and serving it takes time from your event that would be better spent on the dance floor.

- It draws people back to their seats just when they had worked off dinner and started to feel their second winds coming on.

TIP Don't wait until the end of the party to offer coffee. Guests who arrive tired will be thankful for the pick-me-up. If your gathering is a summer barbecue or pool party, put out a pitcher of iced coffee. If it's a dinner, set out coffee by the bar.

At our wedding we served two types of pie, in part because the restaurant specialized in pie. But that's just one of many options to consider. Handheld desserts such as cupcakes, cookies, brownies, donuts, and ice cream cones are great because they're likely to taste better than the cake and, again, they keep people moving around when they're apt to begin losing steam.

No matter what desserts you choose, I recommend they be either passed like hors d'oeuvres or served from a station that guests may approach when they like. The best way to pace courses is to give guests the freedom to pace themselves.

FONDANT? MORE LIKE FONDON'T!

Fondant is the greatest insult to cake since Marie Antoinette. Pastry chefs like it because it's malleable and holds its shape well, so it can be made into elaborate designs that won't fall apart. The results are indeed visually stunning, which is why covering a cake with fondant is the perfect way to impress people who are superficial enough to pass judgment on a food before eating it. In truth, fondant smells like papier-mâché and has the mouthfeel of Big League Chew.

It's no trick to make a beautiful cake out of a substance that's inedible. In fact it's tragic, because there are so many other cake coatings that are not just beautiful but also scrumptious. True Eaters would not sacrifice one iota of deliciousness for aesthetics, and a great host would not expect them to.

THE TIPPLING POINT: WHY FEWER SEATS IS OFTEN BETTER

When hosting a party at home, the principles remain the same: blood flow and pacing are key.

One of the first things to consider is the ratio of available seating to expected guests. The natural inclination is to make sure you have enough spots for everyone, and at the dinner table, that's a necessity. But sometimes fewer seats is better.

When you're hosting a larger party that doesn't include a sit-down meal, I recommend offering less seating, perhaps even hiding chairs that guests may reflexively grab and arrange around the coffee table. That's because of a party phenomenon I've identified as the Tippling Point.

The first few guests usually place themselves on the couch. At a certain point in the party there will be more guests than seats. For a brief period this may feel awkward, as one or two people stand while others sit.

This is the Tippling Point. At this moment, you must resist the temptation to bring in additional chairs.

The Tippling Point
The sooner you get past this awkward moment and get your guests on their feet, the sooner your party will kick into high gear.

Soon more guests will arrive, and those sitting down will get up to mingle. Standing will become the norm, and the couch will fulfill its destiny as a way station. The sooner in your party you pass the Tippling Point, the sooner your guests will begin moving around and the better your party will be.

If your gathering is more intimate, along the lines of a dinner party, seating will be ample, so you need to find creative ways to encourage blood flow. Consider serving hors d'oeuvres and drinks in the kitchen, if your layout allows, so guests remain standing.

TIP Most people either drink too much or not enough. It's your role as host to nudge the teetotalers toward unsobriety and the drunks toward civility.

Whether or not you start on the couch, clear whatever snacks you've put out at least fifteen minutes before moving to the dining table, to allow food to settle and hunger to rebuild.

THE ABILENE PARADOX (OR, HOW TO GET AN INDECISIVE GROUP OF FRIENDS TO GO TO THE RESTAURANT YOU WANT TO GO TO)

The Abilene Paradox is a phenomenon in which a group of people collectively decides to do something that none of them actually wants to do. The term was coined by Jerry B. Harvey and comes from an anecdote in which a family decides to drive to Abilene for dinner. Each person goes along because that's what they think the others want, and they don't want to rock the boat—then they all end up unhappy. In other words, this is what happens when people start being polite and stop being real.

Eater Joshua Wilson of Salem, New Hampshire, called in to *The Sporkful* to offer advice for this type of situation, which can be a sticky one. You're out with a group of people and you're trying to figure out where to go for dinner. You're having trouble agreeing. Some people claim not to care, but then they reject the options others suggest.

Joshua says you can't be too quick to shoot down bad suggestions, in part because you may hurt someone's feelings and in part because if you come across as the person who shoots down everything, you'll lose credibility.

So don't enter the debate too early.

"Let the conversation get to that indecision moment," he explains, "and then you're going to step in and save the day with this magic phrase. And you're going to say this very politely but very decisively. You use the phrase, 'Someone told me about a place.'"

Then you suggest the place you want to go to. Josh likes this approach because it "shifts the responsibility away from everyone in the car" and adds a vague sense of objective authority. But there are pitfalls.

"You can't let the other people realize what's happening, so you have to be really smooth, or they're going to know that you're just trying to manipulate them."

It's a good tip, but don't tell anyone about it—that could render your manipulative powers impotent.

EVENT ATTENDANCE AND THE IDEAL GUEST

No matter how well your host has organized things, it's up to you to act properly as a guest in order to truly maximize deliciousness.

COCKTAIL HOUR: THE BUDDY SYSTEM

When lots of free food and drink are available at once, the frenzy threatens to overwhelm even the seasoned Eater and lead to uncivilized behavior. This is especially true during the cocktail hour, the crucial period that tends to feature the best food of the evening.

Let's assume you're at a pretty ample event, where the cocktail hour features food stations as well as passed hors d'oeuvres. Follow these steps:

1. Get a drink. (See the next section, "Buzz Management," for the reason why.)

2. While drinking, survey the food options as you would at a buffet. Most importantly, FIND THE SOURCE. Figure out where the passed hors d'oeuvres are coming from.

3. Use the buddy system. One person goes to the bar for another round, the other assembles a plate or three of the most promising foods. If your buddy is of the opposite sex, send the guy to get the food, because men are generally shameless and less likely to be judged. (We all know that a man carrying two full plates is just a hungry guy with needs, while a woman with two full plates is an hors d'oeuvre hussy. Plus, the man can take extra food under the guise of chivalry.)

> **TIP** When chatting with other guests, ask them which hors d'oeuvres they like the best. It's a good icebreaker and provides valuable intelligence.

4. Reconvene at a previously determined location, near the source of the passed hors d'oeuvres. This is your base camp.

5. Eat. Drink. Repeat steps three and four as needed. If your base camp is at a prime cocktail table you may want to recruit others to join you, under the pretense of a desire for conversation. They'll hold your table when you and your buddy go foraging.

> **TIP** In the days before the event, ask the host about the event itinerary, in case they have an unorthodox schedule like those recommended earlier in this section. Alter your plan accordingly.

6. Don't save much room for dinner. It's rarely as good.

If the function does not have stations, the basic strategy remains the same, but it becomes even more crucial to find the source of the passed hors d'oeuvres.

BUZZ MANAGEMENT

Managing your intake of food and alcohol is crucial to maximizing your enjoyment of an event. You don't want to get too drunk or too full at any point, but especially not too early in the proceedings. (As Brillat-Savarin said, "Those persons who suffer from indigestion, or who become drunk, are utterly ignorant of the true principles of eating and drinking.")

Food slows alcohol's absorption into your bloodstream. Sometimes that's good, other times it's bad. The key to buzz management is knowing when you want it to happen and when you don't.

In a perfect world, you want your alcohol consumption a little bit ahead of your food intake for most of the night. This gets you a pleasant buzz at the start, then maintains it.

As the night progresses, you can tweak buzz positioning, having a second drink without food or switching to water while nibbling, as the situation warrants. When the end of the affair is in sight, stop drinking before you stop eating, and that last bit of food will help reduce the chance of a hangover. (See figure 10.7, "Buzz Management over Time.")

Let me give you one example of what can go wrong when you employ poor buzz management.

When I attended my friends Allison and Tom's wedding, the hors d'oeuvres were so plentiful that I mistook them for dinner. By the time I ate the actual dinner I was so full that alcohol had no effect on me. I had five glasses of whiskey and was so sober I could have piloted an aircraft, even though I've never piloted one before.

Of course, putting too much booze in your belly without food can have equally dire consequences.

The chart below illustrates optimal buzz management at a large function or party, where you're likely to be eating and drinking over an extended period of time.

BUZZ MANAGEMENT OVER TIME

This represents the ebbs and flows of my ideal wedding buzz experience. You may strive for higher highs, but the basic principle remains the same: Keep alcohol ahead of food, especially early, then alternate between the two to maintain your buzz where you want it. Stop drinking before you stop eating so you can begin to sober up before the event ends, reducing the risk of hangover.

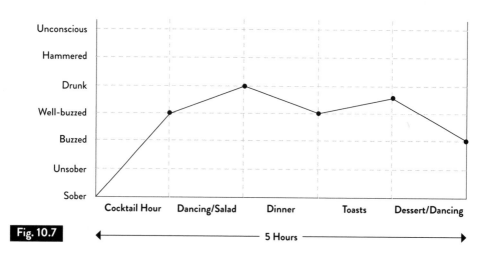

Fig. 10.7 ◄———————— 5 Hours ————————►

DON'T WAIT FOR OTHERS TO GET THEIR FOOD—IT'S IMPOLITE

Mothers everywhere teach that it's impolite to eat your food before everyone has been served. This notion is incorrect. (Fortunately for me, my mother raised me right.)

Even *Emily Post's Etiquette* says, "At a restaurant or an event such as a wedding when you're seated at a large table of eight or more, begin eating once at least three of you have been served."

I'll go a step farther. In the Best Society of the Eatscape, you should start eating

> **TIP** If there are butter and bread on the table when you sit down, reserve some of the butter for later. You may use it to improve subpar side dishes like rice, mashed potatoes, or veggies.

when you get your food, regardless of whether others have been served. That's because as the food gets cold, deliciousness may decrease, and sitting idly by while that happens on the plate before you is the epitome of uncouth.

It's especially unfortunate when some people have their food and those without say, "Go ahead and eat, please, it's getting cold," but one of the people with food insists upon waiting. That person now makes the others with food look bad if they don't wait as well, and everyone suffers.

Allowing food to degrade in quality before you eat it is rude to the chef and shows a fundamental lack of respect for yourself and the greater goals of the Eatscape. It's not good manners.

HOMEWORK

One of the challenging aspects of learning proper etiquette is that not all the rules are universal. At a large function, you want to find the source of passed hors d'oeuvres, but at an intimate dinner party, you wouldn't spend the whole evening standing by the kitchen waiting to accost your hosts whenever they come out with more queso.

When in doubt about how to proceed, take a step back from your setting and consider your role, the number of people there, the venue, how nicely other people seem to be dressed, and whether or not you remembered to put on deodorant.

The answers should point you in the right direction.

If you're still unsure, consult Brillat-Savarin, who said, "He who receives friends and pays no attention to the repast prepared for them is not fit to have friends."

He also said, "Tell me what kind of food you eat, and I will tell you what kind of man you are." Which brings me to your assignment . . .

Write an essay that responds to the following question: Taking into account the preceding two Brillat-Savarin aphorisms, what does someone who receives friends without paying attention to the meal prepared for them eat?

Submit your essay to me at dan@sporkful.com.

COMMENCEMENT

Trustees, deans, faculty, parents, families, and fellow Eaters . . .

Welcome to the first annual Sporkful University Commencement Ceremony. The graduates I imagine before me, as I write this alone in my basement, are a sight to behold. You've worked extremely hard to reach this point, and you deserve a round of applause.

[*Wait for applause to fade.*]

The dictionary defines an Eater as "a person or animal that consumes food in a specified way or of a specified kind." Before reading this book, you probably didn't realize just how many different ways or kinds you could specify. But that, in a nutshell, is the lesson here. Like God, deliciousness is in the details.

By studying and employing concepts such as SATVOR, bite consistency vs. bite variety, the Proximity Effect, ratios within various foods, forkability and toothsinkability, bite force, layering, friction, and palate profit, you have opened up a new world of possibilities each time you put food in your mouth. And by thinking more about language, ethics, cultural awareness, self-actualization, and proper behavior, you've learned to interact with others on this same journey, so that our collective eating experience becomes greater than the sum of its ingredients.

I told you that the road ahead would be arduous—full of challenges, obstacles, risks, and painful realizations. This pursuit of Perfect Deliciousness is a choice, but now that you've made it, you've likely found that it only leads to more choices. How do I hold this sandwich? What do I call this dish? Which shape of this or that food is better? The list goes on so long that you may even wish you could go back to the days when you thought there was only one way to eat something.

You understand now that one of the hardest parts of life as a full-fledged Eater is that you see choices where you saw none before. The stakes at every meal suddenly feel higher, and you may be paralyzed by the fear of making the wrong decision.

I am here to tell you that this fear is misplaced.

For Eaters, the journey is the destination. If you make a bad choice with a bite or a meal, you have not failed. You've experimented. The only failure is not to choose at all. As long as you're thinking, pushing, questioning, innovating, and learning, you're moving closer to the epic grandeur of Perfect Deliciousness.

This life you've chosen isn't easy, but it provides tremendous rewards. Every day when you wake, no matter what else the world has in store, you know that with three meals come three bright spots that illuminate even the most mundane existence. When you start each morning with that knowledge, you'll have a lot more good days than bad.

And when you do, I hope you'll think of your time here at Sporkful University—the lessons you learned, the times we shared—and consider making a donation to our alumni fund.

[*Pause for dramatic effect.*]

Before you throw your caps in the air in celebration, please note that your graduation is not official until you complete the diploma on the following page. Fill in the major of your choosing, cut it out, and hang it in a place of prominence in your home. I recommend the refrigerator.

I'll leave you with one final thought. I can't prepare you for every situation you'll encounter as you head out into the Eatscape. I just hope that this textbook has helped you look at food and eating differently, and that you'll be able to use your newfound knowledge in whatever situations may arise.

That's what Eater Gerome Rothman did. As an American living in Japan, Gerome has dealt with his share of eating difficulties. He e-mailed *The Sporkful* with this anecdote:

> *I recently went to an "Irish Pub" here in Tokyo for some fish and chips. My lunch came with a side of bread and, to my absolute horror, a little pack of margarine to apply to it.*
>
> *Margarine being a crime against humanity with which I will not co-operate, I summoned up my best Japanese to ask for some butter. More margarine arrived. Undeterred, I asked if they had any real butter. Finally, I got a little dish of rock-hard, bread-shredding (but real) butter.*
>
> *I had piping-hot French fries, ice-cold butter, bread, a knife, and little time. I looked at my plate situation and asked myself, "What would the* Sporkful *guy do?"*
>
> *I suddenly had a moment of inspiration. I picked up a French fry and put it on top of the butter. Within 30 seconds, the butter was warm enough to be spreadable.*
>
> *And so the day was saved.*
>
> *I want to thank you for getting me to think more creatively about and have more fun with food. My lunch was not so good, but it was more better. I say that as an English teacher.*

Thank you, Gerome, and thanks to all of you for being such an excellent class. I'll see you around the Eatscape!

DAN PASHMAN
DEAN, SPORKFUL UNIVERSITY

Diploma

The Dean of Sporkful University, with the consent of the Board of Trustees and acting upon the recommendation of the Faculty, has hereby conferred upon

_____
(NAME)

the degree of Bachelor of Eats in _____
(MAJOR)

on this _____ of _____ in the year _____ ,
(DAY)   (MONTH)   (YEAR)

with all the rights, privileges, and deliciousness pertaining thereto. In witness of this action, the seal of the University and the signature of the Dean are affixed herewith.

_____
Dean

SPORKFUL UNIVERSITY

PROMULGATE

RUMINATE

MASTICATE

GLOSSARY

ALPHA . . . a food that is dunked into another food

BACONY BUNDLE OF TWIGS . . . a bunch of asparagus or string beans wrapped in bacon so they can be grilled without falling through the grates

BAGEL TRIFURCATION . . . slicing a bagel horizontally in thirds to improve sandwich ratios; the same technique can be used with muffins (sliced vertically), exposing more interior to buttering and griddling

BALANCING ACT . . . when a larger piece of meat or veggie is balanced on top of a ball of pasta that's wound around a fork

BETA . . . a food that has another food dunked into it

BITE CONSISTENCY . . . when each bite of a food or dish offers essentially the same taste and textural experience as the others

BITE VARIETY . . . when different bites of a food or dish offer notably different experiences from each other

THE BONE SPLITTER . . . a chicken wing consumption technique for two-boned, flat wings where you first remove the smaller of the two bones without removing any meat, to provide unfettered access to succulence

BUBBLING CARBONIC ACID . . . a beverage not unlike water that fizzes but does not sparkle; sometimes incorrectly called sparkling water

BUDDY SYSTEM . . . a cocktail-hour strategy where one person gets drinks while the other gets food and/or scopes out a location for a base camp

BUFFET MASTER . . . a diabolical overlord who seeks to prevent you from getting your money's worth at an all-you-can-eat buffet

THE BULLFROG . . . a popcorn consumption technique whereby you hold a pile of popcorn near your mouth and grab single pieces with your tongue

CAP'N CRUNCH'S COMPLAINT . . . a medical condition characterized by scraping and cutting on the roof of one's mouth

CAPPING (WRAPS) . . . folding in the end or ends of a wrap

CHEWABILITY (GUM) . . . the level of satisfaction produced by chewing a piece of gum

THE CLAW . . . a technique for mixing cereals or snack mixes by using your hand like one of those arcade claw games that dips into a vat of stuffed animals

COMPRESSIVE LOADING . . . biting

CONGRESS OF EQUAL PROPORTIONS . . . when the Eater places an onion ring in the mouth so that its sides align with both sets of molars at once

CONGRESS WITH THE COURTESAN . . . when an Eater eats melon cubed to include a bit of white rind to add a pleasing touch of complementary tart

CRISP (NOUN) . . . like crispiness, only crispier

CRISPER . . . an Eater whose priority when eating a baked potato is maximum exterior potato crisp, even if that means less potato-to-topping contact

CUPCAKE SANDWICHIZATION . . . removing the bottom half of a cupcake and placing it on top, to bring the frosting closer to your tongue and make the cupcake easier to hold and bite

CYCLIC LOADING . . . chewing

DEPRESSION . . . a psychological condition where, after an especially great meal is over, you become deeply sad and withdrawn

DIPPING . . . inserting one food into another to season the food being dipped but not to fundamentally alter it *(example: a tortilla chip dipped in salsa)*

DOMINANT INGREDIENT . . . an overpowering ingredient that easily masks recessive ones

DOUBLE FOLD (POTATO CHIP) . . . a potato chip folded in half twice

DRESSING . . . *(1)* a food placed around (but not inside) another food; a poor substitute for stuffing, or *(2)* sauce or liquid seasoning for salad

THE DRIZZLER . . . pouring milk over cereal in a circular motion to ensure that all cereal is milked

DUNKING . . . inserting one food into another such that at least one of the two foods is irreversibly changed *(example: a donut dunked in coffee)*

EATABILITY . . . the compatibility of a food with the Eater's interface

EATCONOMY . . . the wealth and resources of your stomach, as well as the flow of goods through your stomach

EATER . . . a seeker of the Platonic ideal known as Perfect Deliciousness

EATER ACTUALIZATION . . . a psychological state embodied by an Eater who exhibits morality, creativity, spontaneity, problem solving, lack of prejudice, and acceptance of facts; the highest level in Maslow's Hierarchy of Needs

EATSCAPE . . . a community brought together by its members' common passion for eating and seeking deliciousness

EGG BEATER TECHNIQUE . . . replicating the motion you use to beat eggs in order to mix cereal and milk, to cover all the cereal with milk when you've added the milk first

EGGTEGRITY . . . structural integrity in an egg-based food

EGO . . . the part of the stomach between the id and superego, where the stomach's fill line is located; often rationalizes increased consumption

EL MIXTEC . . . mixing all the fillings of a wrap or burrito before wrapping them in the tortilla or wrap

EXTREME PICNICKING . . . an elaborate picnic setup with multiple courses, plates and utensils, and cloth napkins; usually only happens in movies and catalogs

EYE OF THE STORM (POTATO CHIP) . . . a potato chip that's smaller than its brethren, so it ends up cooked medium well, crunchy, and charred

FACE FUNNEL . . . a popcorn consumption technique whereby you hold a handful of popcorn against your face and let it cascade into your mouth

FALLINGSANDWICH . . . a Buffalo chicken sandwich constructed as a tribute to Frank Lloyd Wright's Fallingwater

FIXINS . . . the accoutrements traditionally served with a food, such as lettuce and tomato with a burger, or Israeli salad and hummus with falafel

FORKABILITY . . . the ease with which a food can be stabbed onto a fork and kept there until eaten

FORKED TONGUE . . . a chicken wing consumption technique for two-boned, flat wings where you nibble the exterior, then use your tongue to poke out the meat between the bones

FRIED . . . the crispy exterior of a fried food, as in "I wish this fried chicken had more fried."

GENERALIZED ANXIETY DISORDER . . . a psychological condition that causes you to worry constantly that your food won't be delicious enough and/or that there won't be enough of it

HERETIC'S BUFFET . . . a family of dishes that combine traditional Jewish foods with pork and/or shellfish

THE HIGH HORSE . . . a technique for wrapping a wrap where all fillings are piled in the middle and the wrap is folded over them

HORIZONTAL CONGRESS . . . when an Eater eats melon squares, rectangles, or trapezoids cut to provide flat sides that lie pleasantly flush against the teeth

HUMMINGBIRD TECHNIQUE . . . getting seconds at a buffet without waiting in line again by hovering and darting into openings; also known as the Hover-and-Dart

HYPOCHONDRIA . . . a psychological disorder in which you always think you're hungry, even when you're not

ICE COFFEE . . . iced coffee made with coffee ice cubes

ID . . . the part of the stomach that growls

INITIAL BITE AREA . . . the region of a food from which the Eater may choose the first bite

INVERTED SALAD . . . a salad with the greens on top and more substantial components underneath

IN VITRO POPCORN-BUTTER FERTILIZATION . . . a method for jury-rigging a movie theater butter dispenser with straws to ensure even butter distribution throughout popcorn

IRONING OF THE SHEETS . . . when an Eater presses down on the top of a very tall sandwich to flatten the layers and make it mouth-ready

JACK SPRAT COROLLARY . . . the notion that if you're going to remove a portion of a food, and leave only a part that may be unappetizing to others, you have to ensure there's someone present with complementary tastes who will eat the part of that food that you left

KAMA SUTRA OF CONSUMPTION . . . a series of positions meant to maximize pleasure during the eating act

LEAF THICKNESS UNIT (LTU) . . . a unit of measurement equal to the thickness of one leaf of lettuce or comparable salad green

LEGIBILITY (SALAD) . . . the ease with which a salad can be transported to the mouth

LETTUCE GLOVE . . . a leaf of lettuce used to grip a food or group of fillings in a sandwichesque arrangement

LIVING SANDWICHDOM . . . a school of thought that argues that sandwiches are dynamic foods that change over time, so that the definition of a sandwich should not be restricted by the Earl of Sandwich's intentions when he created this food centuries ago; opposite of the Strict Sandwich Constructionism approach

MARRIAGE OF CONVENIENCE . . . when an Eater eats melons cut into large wedges with the rind attached, making them easy to hold in the hand but messy on the face to eat

MARTYR . . . a host who sacrifices his/her own enjoyment so that guests may be happy

THE MEAT UMBRELLA . . . a chicken wing consumption technique for two-boned, flat wings where you stand a wing on a plate, smaller end up, and press downward along the bones from all sides, causing the meat to strip off the bone and turn the wing into the shape of an umbrella

MÉNAGE À 3.14159 . . . a French term for the area of pie where the bottom outer crust, vertical side crust, and fillings all come together

THE MID-FINGER PINCER . . . a technique for picking up a beverage when your hands are very messy that involves gripping the glass with the pads in the middle of the fingers

MISSIONARY . . . a host who leads by example, setting a tone of revelry and merriment that becomes contagious

MISTLEHOCK . . . a superior alternative to mistletoe made by hanging a ham hock from the ceiling

MULTIPLE PERSONALITY DISORDER . . . a psychological disorder that leads you to assume multiple personalities when free food is available, in order to take as much as possible without seeming greedy

MUTUAL TRANSFERENCE . . . when one food is dunked into another and both are changed as a result

NEW YEAR'S LETDOWN SYNDROME . . . the inevitable reaction when overhyped New Year's Eve plans fail to meet expectations

OBSESSIVE-COMPULSIVE DISORDER . . . a psychological condition in which you compulsively turn on the stove to cook things, taste food repeatedly, and wash your hands after eating

ONE HAND, TWO CHIPS RULE OF NACHO MORALITY . . . the notion that when taking nachos from a pile, you should only use one hand and only grab up to two chips at once with that hand

ONION DONUT . . . a really, really heavily battered onion ring

ORIGIN OF SPICES . . . a theory that states that an appreciation for spiciness that's complementary without being dominant is a sign of a more evolved Eater

THE PALM . . . when you use your palms to pick up a beverage because your hands are very messy

PANIC DISORDER . . . a psychological condition in which you suffer from recurrent panic attacks, perhaps resulting from agoraphobia (the fear of being in a public place without enough food) or claustrophobia (the fear of being in a confined space without enough food)

PERFECT DELICIOUSNESS . . . a gustatory nirvana, a higher state, greater even than the sum of all the sensory pleasures, derived from eternally consuming the ideal bite

PIPELINE (POTATO CHIP) . . . a potato chip that's folded over but not flat, so it resembles a wave about to crash

POPCORN MOMENT . . . an instance of eating discovery or inspiration

POPPING THE PLUG . . . a technique for chewing capped pens in which you use your teeth to remove the plug at the back end, chew it until it's folded over, then reinsert it sideways

PORKLIFT . . . a bacon lattice structure that elevates a pancake stack off the plate, so the bottom pancake doesn't become soggy with syrup

POTLUCK DINNER . . . a meal in which all participants bring food but none knows what the others are bringing, so the success of the meal is largely determined by luck

PROXIMITY EFFECT . . . the notion that when you put any multifaceted food into your mouth, the components in closest proximity to your tongue are the ones you'll taste the most

PRUNING THE HEDGES . . . when an Eater nibbles the perimeter of a sandwich to trim fillings that protrude beyond the bread boundary

THE **R**AKE'S PROGRESS . . . a chicken wing consumption technique for two-boned, flat wings where you snap one of the end joints, spread the bones into a V shape, put the whole thing in your mouth, and pull it out while using your teeth to rake off the meat

RECESSIVE INGREDIENT . . . a mild ingredient in danger of being overpowered by a dominant one

REFRIGERATOR BLINDNESS SYNDROME . . . a condition that prevents people from seeing some foods in a refrigerator; afflicts people with gaps in their mastery of object permanence

RIDING THE WAVE . . . getting seconds at a buffet without waiting in line again by monitoring ebbs and flows and bottlenecks, then striking where there are breaks in the crowd waters

SAFE WORD . . . when an Eater arranges the hands like a net around the back of a sandwich to restrain disobedient fillings

SALSA SLALOM . . . maneuvering a chip through multiple dips in a slalom pattern, to avoid vitiating one dip with another

SALTWATER SPRINKLE . . . a method for salting a hard-boiled egg that's rooted in the Jewish observance of Passover but useful for all people, all year round

SANDWICH . . . a food that satisfies two basic criteria: 1) you can pick it up and eat it without your hands touching the fillings, and 2) the fillings are sandwiched between two separate, hand-ready food items

SANDWICH GENOME PROJECT . . . a worldwide effort to map the genetic makeup of every sandwich in existence

SATVOR . . . surface-area-to-volume ratio, essentially the ratio between a food's volume and the amount of exterior it has exposed to its surroundings; affects a food's temperature, texture, flavor, and much more

SAUCEABILITY . . . the degree to which a food keeps sauce adhered to it

SAUSAGE GRAVY APEX . . . the physical and emotional high that results from the first bite of sausage gravy, making such a strong impression that it drives you to order sausage gravy again in the future, despite the fact that the Apex is quickly followed by sadness

SEIZING THE SPARROW . . . when an Eater is at the halfway point of eating a coated ice cream pop and skips ahead to the bottom two corners, because they're the two best bites and they've reached the perfect level of meltiness at that precise moment

SEMOLINA FULCRUM . . . the point at which the relationship between sandwich bread hardness and filling stability becomes structurally sustainable

SHOVEL FORMATION . . . an arrangement of short pastas on the fork such that they're perpendicular to the tines, forming a type of shovel that can be used to scoop sauce and other ingredients

SIDECAR . . . an espresso cup or other small vessel used to hold liquid for the purpose of dunking a food into it without leaving crumbs or other residue in your actual beverage

SINGLE FOLD (POTATO CHIP) . . . a potato chip folded in half once

THE SINGLE STREAM . . . pouring milk into cereal in a single spot, so some surface cereal stays dry

SLICEABILITY . . . the level of ease in slicing a food without disrupting its structural integrity

SLICED AVOCADO GORDIAN KNOT . . . the problem posed by sliced avocado in a sandwich, namely that if not secured properly, it tends to slide out the back when confronted with bite force

SLICED CUCUMBER CONUNDRUM . . . the problem posed by sliced cucumber in a sandwich, namely that if not secured properly, it tends to slide out the back when confronted with bite force

SLICED TOMATO BOTHERATION . . . the problem posed by sliced tomato in a sandwich, namely that if not secured properly, it tends to slide out the back when confronted with bite force

SLIDEAGE . . . the condition of sliding; similar to slippage

SLIPPAGE . . . the condition of slipping; similar to slideage

THE SNAIL'S COCHLEA . . . a technique for wrapping a wrap where fillings are spread out on the wrap and it's wrapped in small, tight circles, so a cross section produces a spiral

SOAPELLIER . . . an expert in scented soaps and the art of pairing such soaps with different foods

SOGGABILITY . . . the likelihood of a food to experience soggage

SOGGAGE . . . like sogginess, only more demoralizing

SPLITTER'S DILEMMA . . . a quandary whereby the more you open a baked potato to expose its interior to delicious toppings, the more you also flatten its skin against the plate and cause condensation, eating away at exterior crisp

STAINABILITY QUOTIENT . . . the probability that a food or serving method will result in you staining your clothes, or someone else's

STALKING THE SIREN . . . when an Eater eats melons cut into balls, which possess a certain aesthetic beauty but are difficult to keep on a plate

STARBURST (POTATO CHIP) . . . a potato chip that resembles a starburst or other interstellar phenomenon

STEAK SATIETY PARADOX . . . the notion that the more delicious steak you eat, the hungrier you become

STIFFNESS . . . a measure of structural rigidity, determined by the degree to which a sandwich bread has been toasted or griddle-grilled before sandwichization

STRENGTH . . . a sandwich bread's ability to support the sandwich mass, especially in tension from compressive loading (biting) and cyclic loading (chewing)

STRICT SANDWICH CONSTRUCTIONISM . . . a school of thought that argues that we must look only at the Earl of Sandwich's original intent in creating his eponymous masterpiece in order to find the limits of sandwichdom; opposite of the Living Sandwichdom approach

STUFFING . . . a food placed inside another food; its primary purpose is to absorb flavor from its host, but it may impart flavor as well

SUPEREGO . . . the part of the stomach that feels revulsion at the thought of excessive eating, and produces pangs of guilt (also called "nausea") if you do it anyway

SWALLOW'S CIRCUMSTANCE . . . when an Eater eats melons cut into very narrow wedges, so they're less substantial but can fit in the mouth and be eaten by hand without facial mess

TART (noun) . . . like tartness, but without the negative connotation

TEMPERATURE OF LIFE . . . the precise temperature of the air in the Eater's immediate vicinity at any moment; a scientifically superior alternative to "room temperature"

TIME . . . a seasoning to be applied to foods, generally by letting them sit while time elapses

TIPPLING POINT . . . the moment during a large party in a home, as guests arrive, when there is one more guest than there are seats; the sooner the party passes this point and seated guests stand up, the sooner the party will kick into high gear

TOOTHPICNIC . . . a picnic where all foods are served in bite-size pieces, so they can all be eaten with toothpicks, eliminating the need for plates, utensils, and most napkins

TOOTHSINKABILITY . . . the amount of satisfaction an Eater receives from sinking his/her teeth into a food

TOPPABILITY . . . the ease with which toppings can be added to and kept on a food

TOPPER . . . an Eater whose priority when eating a baked potato is maximum topping-to-potato contact, even if that means more skin is pressed against the plate, which causes condensation and reduces crisp

TWINING OF THE CREEPER . . . when an Eater wraps food such as bacon strips or string cheese strands around the finger with the desire of kissing it

UNREQUITED TRANSFERENCE . . . when one food is dunked into another but only one is changed as a result

VEGGIEDUCKEN . . . a large, time-consuming, centerpiece-worthy, vegetarian dish fit for major holiday events; made of sweet potatoes inside leeks inside a giant squash, with stuffing between the layers

"WATERMELON" . . . artificial watermelon flavoring

WATSON AND CRICK . . . a maneuver for getting short, cylindrical pastas on the fork that involves sliding the tines through the pasta's hollow cavities

WEBER'S LAW . . . the notion that a grill generates heat, people talking about grilling generates more heat (in the form of hot air), and as more people grill and talk about how great grilling is, grilling grows more popular (thus changing the system), and leading to more grilling (thus increasing work); an example of the First Law of Thermodynamics

WINDOW OF OPTIMAL CONSUMPTION . . . the period of time when a food's changing characteristics are at their most delicious

ACKNOWLEDGMENTS

So there you have it, friends, the first textbook in the history of Sporkful University, and my first book of any kind. Whew! Now I know how Philip Roth feels.

In many ways this book is the culmination of five years working on *The Sporkful* and twenty years working as a writer, reporter, producer, editor, host, waiter, cab driver, tennis instructor, and Bioterrorism Project Coordinator (seriously). So I have a lot of people to thank. Let's begin at the beginning.

My family and friends have always been incredibly supportive in all respects, and especially throughout my many career ups and downs. Thank you Mom, Dad, Howard, Manya, Grandma, Alice, Gene, Danny, Beth, Gabriel, Noa, and all my aunts, uncles, cousins and friends far and wide, including my cousins Eric Meller and Lori Hoffman, who designed the *Sporkful* logo and the first two *Sporkful* websites. (By the way, if you liked this book, check out my brother Howard's dissertation. It's called *Making Revolution Work: Law and Politics in New York, 1776–1783*. He's a legal historian.)

Special thanks to my daughters, Becky and Emily, and to my amazing wife, Janie, a.k.a. "Mrs. Sporkful." She didn't just put up with my many late nights and weekends spent working on *The Sporkful* in its early days, she encouraged it and made it possible, even after I did unspeakable things to our waffle maker. I love you, Janie!

Now let's talk about the people who made this book. *Eat More Better* exists because Kate Lee at ICM and Mike Szczerban at Simon & Schuster understood it and got excited about it. I owe them both a lot. Mike was my editor throughout the writing process and was an ideal creative partner, guiding the book masterfully even though at some points neither of us was exactly sure where we were going. Sarah Knight at S&S took over for Mike and made the transition seamlessly. She was crucial in turning the vision into a reality. Meg Cassidy and Andrea DeWerd took on the rather difficult task of convincing people to buy a book that, I'd like to think, isn't quite like other food books, and they did it with enthusiasm and expertise. Jenn Joel and Kari Stuart at ICM have been extraordinarily supportive of and patient with a first-time author who asks a lot of questions. They've never steered me wrong.

Alex Eben Meyer's illustrations add a whole other dimension to this book. Time after time I gave him a description of a straightforward illustration and he came back with something that blew my mind. Thanks, Alex!

Thanks to Mark Garrison, whose enthusiasm and dedication were crucial in *The Sporkful*'s early days. His creativity and wit inform so much of this book. Thank you, Mark.

Thank you also to Jon Karp, Jonathan Evans, Jason Heuer, Marlyn Dantes (who designed the awesome cover), Shade Grant, James Gregorio, Becky Cole, and everyone at S&S, ICM, and elsewhere who I failed to mention but who contributed in ways large and small.

Thank you to the Eaters all over the world who acted as recipe testers and fact checkers for this book: Caitlin Abram, Michele Anderson, Summer Ash, Van and Harrison Brenner, Todd Chandler, Craig, Gayleen, Jude & Theo Dimond-Bauer, Tom, Allison, Jonathan and Julia Ehlers, David and Ethan Emmert, Michael Foody, Nathaniel Goodyear, Adam, Stacey, Noah and Zachary Gould, Ryan Hennessy, Jonathan Jenkins Ichikawa, Carrie Ichikawa Jenkins, Stephanie

Jones, James H. & Yuka Lui, John and Linda Lyden, Claire Morgan, Joshua Pesikoff, Steve Phifer, Stacey Privia, Alison, Owen and Emmett Ruggeri, Laura, David, Alex and Annabel Schultz, Oliver Toothaker, Rachel Turner, Laura and Zoë Wynohrad, Steven Yates, and Stephanie Young.

Then there are the many smart, talented, thoughtful, passionate people I've been fortunate enough to work with over my many years on various creative projects. Their skill and dedication have made me better at what I do, and remind me that I always have more to learn. There's a piece of each of them in this book. Listing them here may seem indulgent, but what is this book if not indulgent? So with sincere apologies to those I've accidentally omitted, thank you: Rupert Allman, Isaac Aronson, Laura "L.V." Anderson, Jesse Baker, Chris Bannon, Zena Barakat, Chris Benderev, Alex Blumberg, Andy Bowers, Luke Burbank, Madeleine Brand, Dean Cappello, Lucy Carrigan, Zoe Chace, Bruce Cherry, Ian Chillag, Laura Conaway, Jay Cowit, John Crimmings, Mike Danforth, Nathan Deuel, Melissa Eagen, Jim Earl, Amy Eddings, Tim Einenkel, Angela Ellis, Robin Epstein (who said, "Don't call it *The Spork*, call it *The Sporkful*"), Amy Epstein Feldman, David Folkenflik, Jacob Ganz, Alex Goldmark, Sarah Goodyear, Liza de Guia, Arwa Gunja, Vicki Hippel, John Hockenberry, Sharon Hoffman, Mike Hogan, Dan Holloway, Taige Jensen, Kent Jones, Caitlin Kenney, Jay Kernis, Lauren Kirchner, Robert Krulwich, Ron Kuby, Barry Lank, Katherine Lanpher, Jim Ledbetter, Shelley Lewis, Julia Lipkins, Alex Lisowski, Leonard Lopate, Kris LoPresto, Rachel Maddow, Fritz Manger, Marion Maneker, Laurie March, Marc Maron, Ilya Marritz, Rachel Martin, Matt Martinez, Brendan McDonald, Ben McGrath, Trish McKinney, Kristen Meinzer, Stephen Metcalf, Kristen Muller, Tina Nole,

Eric Nuzum, Sarah Oliver, Max Osswald, Mike Pesca, Alex Picarillo, David Plotz, Deb Puchalla, Katie Quinn, Nazanin Rafsanjani, Jack Rice, Josh Rogosin, Chris Rosen, Win Rosenfeld, Jen Rosta, Peter Sagal, Adam Silver, Laura Silver, Vanessa Silverton-Peel, Mike Singer, Bill Smee, Margaret Low Smith, Robert Smith, Lauren Spohrer, Dana Stevens, Alison Stewart, John Swansburg, June Thomas, Julia Turner, Krishnan Vasudevan, Chris Vivion, Andrew Walsh, Joy Wang, Manoli Wetherell, Jason Williams, Lizz Winstead, and Robin Young.

INDEX

beer, 20, 43, 173
  biology and, 269–70
  craft, 140–42
  economics and, 133, 140–43
  marketing gimmicks of, 142–43
  philosophy and, 101–2
belt loosening and tightening, 146
beverages:
  biology and, 249, 254, 266–70
  cultural studies and, 156–57, 162,
    173, 175–76, 184
  economics and, 131, 133, 135–44
  etiquette and, 276, 278–79, 284–86,
    288, 290–95
  language arts and, 36, 41, 43, 50–52,
    65–66
  mathematics and, 196–97
  philosophy and, 101–2
  physical sciences and, 13–15, 17–20,
    30–32
  psychology and, 222–23, 226, 229,
    235–36, 238
bibs, 277
biodiversity, 244, 250–51
bioengineering, 251, 263–72
  piña colada reconstituted and,
    268–69
  Popcorn Moment and, 263–64
  salamis and, 265–68
  Sporkful University and, 269–70
biology, 243–72, 274
  environment and evolution in,
    245–51, 255
  genetics and taxonomy in, 251–63
bird trawlers, 134–35
birthdays, 181, 205, 222, 287
bite consistency, 2, 148, 298
  engineering and, 70, 75–77, 82, 85–87
  philosophy and, 111–12
bites, 1–2, 144
  biology and, 245, 258–60, 262

  commencement and, 298–99
  cultural studies and, 162–63, 184
  engineering and, 70–71, 73, 75–76,
    78–86, 89, 94
  language arts and, 49, 55–58, 66
  mathematics and, 190, 192–93, 202,
    209
  philosophy and, 101, 114–15,
    118–21
  physical sciences and, 21, 26–28, 32
  psychology and, 227–28, 230, 232,
    239
bite variety, 2, 112, 298
  economics and, 148–49
  engineering and, 76, 86–87
blue cheese dressing, 70, 84
blue crabs, 30
Bone Splitter, The, 246
Brazil Nut Effect, 148–49, 250–51
bread crumbs, 167–68
breads, 294, 300
  biology and, 253, 262, 272
  crustiness of, 72–73, 78–79
  cultural studies and, 156, 159, 162,
    170–73
  defrosted, 12
  economics and, 127, 135, 142, 144
  engineering and, 70–74, 77–84, 88,
    91
  grilling of, 16–17
  language arts and, 37–41, 49, 52
  mathematics and, 192, 198–201
  philosophy and, 99–100, 102–6, 118
  physical sciences and, 9, 11–13,
    16–17, 22, 26
  popular sandwich, 78–79
  psychology and, 224–25, 237–38,
    241–42
  sliced and slicing of, 3, 11–13, 78–79
  thickness of, 78–79, 198
  see also toast, toasting

breakfasts, 96, 135
   cereal combos as, 150
   cultural studies and, 160, 172
   philosophy and, 99–100, 110
breath maintenance, 279–81
Brillat-Savarin, Jean Anthelme, 25, 293, 296
brining, 157–58
brisket, Texas vs. Jewish, 255
brunches, 99, 132, 141
buddy system, 291–92
Bud Light, 142–43
Buffalo chicken sandwich, 70, 84
Buffalo sauce, 41, 249
Buffalo wings, 36
   biology and, 244–50
   The Bone Splitter technique for, 246
   evolved theory on, 248–50
   The Forked Tongue technique for, 248–49
   The Meat Umbrella move for, 247
   The Rake's Progress maneuver for, 246–47
Buffet Master, 64, 120
   defeating of, 127–28
   economics and, 126–31
buffets, 286, 292
   all-you-can-eat, 3, 62–64, 126–33
   cultural studies and, 156, 171–72
   economics and, 126–33, 143
   foods that call for caution at, 132
   game theory and, 128–31
   The Hummingbird Technique at, 130–31
   language arts and, 62–64
   Riding the Wave at, 130–31
   sunk costs and, 132
   typical setup of, 129
Bullfrog, The, 264
bullying, 231
buns, 26, 39, 144

engineering and, 72, 82
   grilling of, 16–17
   psychology and, 224–25
   sizes of, 201–2
burgers, 17, 84
   cultural studies and, 166, 183–84
   language arts and, 42–43
   techniques for dipping of, 230
   see also cheeseburgers
burritos, 90
   biology and, 255–56
   language arts and, 38–39
business, see economics
butter, 17, 45, 165, 294, 300
   biology and, 264–65, 271
   economics and, 132, 135, 138, 144
   engineering and, 77, 79, 94–95
   mathematics and, 201, 203
   philosophy and, 103–4
   psychology and, 225, 231, 242
buzz management, 4, 274, 293–95

C

Caesar salads, 49, 54–55
cakes, 83, 147, 241, 255
   cultural studies and, 154, 181–82
   etiquette and, 287–88
   literary arts and, 44–47
   philosophy and, 99–100
calzones, 263
candy, 53
   approach to consumption of, 216
   cultural studies and, 165, 179
   psychology and, 215–19
Cap'n Crunch's Complaint, 59, 79, 195
Caramel deLites, 44–45
carbon dioxide, 50–51
carnival-style corn poppers, 265
cars, eating in, 154, 182–84
casseroles, 145, 163, 166
cavatappi, 257–58

commencement, 297–300
competitive eating, 156, 245–46
concerts, 136–37
condensation, 21–24
    baked potato toppings and, 23–24
    perils of, 21–22
condiments, 40, 135, 202, 267, 282
    cultural studies and, 182–84
    engineering and, 71–72, 79, 84–85
    psychology and, 225, 229–30
confidence, 224–25, 227, 234, 242
Conti, Crazy Legs, 245, 247
cookies, 28, 102, 180
    language arts and, 43–47
    psychology and, 235–36
cooking, 64, 108
    biology and, 255, 260–61, 263, 266–67,
        271–72
    cultural studies and, 157–62
    economics and, 132–33, 145
    engineering and, 89, 92–93
    etiquette and, 283–84, 295
    mathematics and, 195, 208
    physical sciences and, 15–17, 21, 26
    psychology and, 234, 237
Coors Light, 142
Copernican Heresy S'mores, 27
cost-benefit analysis, 126, 137–44
Crab Nebula, 30
crackers, 26–28, 100, 102–3
cream, creaminess, 18, 237, 269
    cultural studies and, 168–69, 171
cream cheese, 79, 171
    language arts and, 40–41, 46, 50
crispness, 39, 132
    biology and, 246, 249, 262–63
    cultural studies and, 160, 167–68,
        170–71, 183
    engineering and, 70, 86, 92, 94–95
    mathematics and, 195, 201, 207
    philosophy and, 101, 105

physical sciences and, 8, 10, 17,
    21–24, 33
    psychology and, 231–32, 237, 240–42
    vertical sandwich plating and, 22
Crispy Salad Poppers, 59
crostini, 171–72, 242
croutons, 54
crumbs, 28, 103–4, 183, 235
crunchiness, 119, 174, 251
    economics and, 134–35, 148, 150–51
    engineering and, 79, 86, 94
    mathematics and, 194–95, 206–7
    physical sciences and, 8, 21, 33
crustiness, 255
    language arts and, 44–45
    mathematics and, 189–91, 198–99, 204
    philosophy and, 101, 105
    psychology and, 237, 241
    see also pies, pie crusts; under
        breads
cube farms, 177–78, 180–81
cucumbers, 74–75
cultural studies, 153–85, 298
    cakes and, 154, 181–82
    and eating at work, 154, 177–82
    and eating in the car, 154, 182–84
    holidays and, 154–76, 185
    and mac and cheese, 166–70
cupcakes, 83, 288

D

Danforth, Mike, 39–40
Darwin, Charles, 248–49, 272
Death Stars, 29
deliciousness, 298
    biology and, 244, 246, 250, 255–56,
        260–61, 272
    cultural studies and, 154, 157, 170,
        177, 184
    economics and, 126, 128, 137, 140–41,
        146, 148–49

evolution, 245–51, 255
    Buffalo wings and, 248–50
exponents, 188, 204–7
Eye of the Storm chips, 195

F

Face Funnel, The, 264
factorials, 208
fad diets, 143–44
Fallingsandwich, 70, 84
fast foods, 99, 183
Fibonacci, Fibonacci sequence, 202–4
fillings, 28, 283
    engineering and, 70, 73–74, 76, 78–81,
        83, 85–88, 90
    language arts and, 37–38, 46
    mathematics and, 189–91, 198–99
    philosophy and, 105–6
first dates, 227–28
fish, 42, 54, 78, 99, 207, 252, 295
    cultural studies and, 174, 177
    economics and, 132, 134–35, 138,
        145
Fitzgerald, Kellie, 283
fixins, 84
flagels, 199
flavors, 283
    biology and, 246, 250–51, 253–54
    cultural studies and, 159, 168–69,
        174
    economics and, 135–36, 147, 150
    engineering and, 79, 84, 90, 92
    language arts and, 40–41, 53–54, 64
    mathematics and, 190, 194–95,
        197–98, 202–3, 207, 210
    philosophy and, 101–2, 104
    physical sciences and, 15, 17, 31
    psychology and, 225, 231–32, 235,
        238, 241
Foil and Water Bath Technique, 46
food festivals, 122–23

foods:
    expensive and cheap, 144
    moving between drinks and, 278–79
    naming of, 37–44
    sandwichesque, 83–84
    sharing of, 227–28
    words that make them sound
        superior, 49
food temperature, 15–24, 26–27, 31
Forked Tongue, The, 248–49
forks, forkability, 3, 31, 76, 167, 207, 298
    biology and, 257–61
    language arts and, 55–58
    psychology and, 227–28
fractions, 188, 198–204
French, John, 277–78
French fries, 21, 99, 300
    bullying and alternative-lifestyle
        dips for, 231
    cultural studies and, 154, 182–84
    psychology and, 231–32
French Fry Napoléon, 231–32
Freud, Sigmund, 214–15, 219, 226
fried chicken, 286
    biology and, 262–63
    language arts and, 37–38
    SATVOR and, 8–10
fried foods, 132, 207
    biology and, 249, 255, 262–63
    cultural studies and, 154, 160, 162,
        165–66, 170–71, 178
    physical sciences and, 8–10, 21
    psychology and, 231–32
    SATVOR of, 2, 8–10
frostings, 41, 83, 181
frozen yogurt, 42–43
fruits, 135
    cultural studies and, 174–76, 179
    language arts and, 53–54, 59
    mathematics and, 196–97, 203–4
    physical sciences and, 21, 28, 32

Kit Kats, 216–19
knives, 2, 94, 103, 228, 300
    cultural studies and, 165, 177
    language arts and, 54–56
    mathematics and, 205, 209
    physical sciences and, 12, 31
Konn, Emily, 45
kosher foods, 134, 266
Krulwich, Robert, 192–93

L

Lablans, Mirjam, 161
lamb, 131–32
language arts, 35–67, 274, 298
    Girl Scout cookies and, 43–48
    grammar and usage in, 53–59
    misnomers in, 49–53
    oatmeal and, 48
    poetry and composition in, 60–67
    regionalism in, 41–42
    semantics and etymology in, 37–53
    and words that make foods sound
        superior, 49
    writer's tools and, 64
large movie theater corn poppers, 265
lasagna, 261
latkes (potato pancakes), 37, 170–72
Leaf Thickness Units (LTUs), 75
Leftover Ham Sandwich with Mac and
    Cheese Spread, 170
leftovers, 223
    cultural studies and, 159–60, 164, 170
    economics and, 128, 145
    Thanksgiving and, 159–60
lettuce, 2, 54, 172, 209, 286
    engineering and, 73, 75, 84–85
Lettuce Glove Technique, 84
Levin, Michael and Carla, 266
liquids, 267
    cultural studies and, 159, 162, 170
    engineering and, 70–73, 79, 82, 94

gravy and, 79, 120–21, 231, 233
    physical sciences and, 9–10
    psychology and, 231, 233, 235, 241
lobster, 30, 144, 169, 277
logic, 210
loss leaders, 141
love:
    psychology and, 224–25, 227–38, 242
    sausage gravy consumption and, 233
lox, 79, 99, 171
lunches, 99, 166, 177, 300

M

mac and cheese, 286
    ideal preparation of, 167
    innovations with, 169–70
    recipes for, 168–70
McDonald, Brendan, 248–49
Maddow, Rachel, 197, 268–69
Maddow Colada, The, 269
Makeshift Vehicular French Fry
    Dipping Basin, 184
Mamet, David, 54
Manischewitz Sorbet, 176
Manischewitz wine, 175–76
marinara sauce, 215, 231
market inefficiencies, 126, 145
marketing gimmicks, 142–43
maror, 175
marshmallows, 26–27
Mashed-Potato-and-Gravy Vessel, 120–21
mashed potatoes, 120–21, 159, 231–32,
    294
Maslow, Abraham, 215, 224–25, 227,
    231, 234, 239
mathematics, 187–211
    algebra and advanced, 207–9
    exponents and, 188, 204–7
    geometry and, 188–96
    and ratios and fractions, 188, 198–204
matzoh, 172–74

multiple personality disorder, 234
Murray's Cheese, 228
mustard, 120, 168, 238
    biology and, 252–53, 267
    language arts and, 40–41

N

nachos, 49, 138, 211, 229, 239
napkins, 31
    biology and, 246, 249
    cultural studies and, 183–84
    etiquette and, 275–78, 283
Newton, Isaac, 25, 71–77
New Year's Eve, 155, 162–63
New York Times, 133
"Nice Cup of Tea, A" (Orwell), 65
nuts, 148
    cultural studies and, 175–76, 179
    language arts and, 54, 56

O

oatmeal, 48, 100
object permanence, 223
obsessive-compulsive disorder, 234
oils, 105, 242, 265
    cultural studies and, 157–58, 168–69,
       172, 178
    engineering and, 71–73, 79, 90
    mathematics and, 200, 209
olive oil, 200, 242
    cultural studies and, 157, 178
    engineering and, 72, 79
Oltman, Tim, 168
Omega-3^3, 207
omelets, 43, 99, 178, 206
    engineering and, 88–92
    folding of, 90–91
    as inside-out-sandwich, 91
    ranking different ways of spelling, 92
One Hand, Two Chips Rule of Nacho
    Morality, 239

onion donuts, 52
onion rings, 52, 119
onions, 84, 252
    cultural studies and, 166, 178, 183
    fried, 166, 178, 207, 231–32
    psychology and, 231–32
oral fixation, 219–22
oranges, 32, 101, 204
order of operations, 208–9
Oreos, 180
Orwell, George, 65–66
oxymorons, 42–43

P

Paleo Diet, 144
Palm, The, 279
pancakes, 4
    engineering and, 70, 95
    potato, 37, 170–72
    waffles vs., 95
panic disorder, 234
Passover, 255
    cultural studies and, 170, 172–76
    enjoyment of, 173–76
Passover Sangria, 175–76
pastas:
    biology and, 254–55, 257–63
    classes of, 258–62
    controlling portioning of, 259
    cultural studies and, 166–70, 173, 177
    economics and, 134–35
    language arts and, 41, 49
    long, 258–60
    phyllum criteria of, 257–58
    shapes of, 3, 257–58, 262
    short, 260–62
pastrami, 241
peanut butter, 28, 100, 172, 193
    engineering and, 78, 92–93
    language arts and, 40, 44–47
    waffles with sealant of, 93

ABOUT THE AUTHOR

Dan Pashman is the creator and host of WNYC's James Beard Award-nominated podcast *The Sporkful*. He also hosts the Cooking Channel web series *Good to Know* and *You're Eating It Wrong* and is a contributor to NPR, *Slate*, *Buzzfeed*, and LA's KCRW. He lives with his wife and two daughters outside New York City.

Visit Sporkful.com to listen to *The Sporkful* podcast, watch some videos, and say hello!

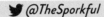 *@TheSporkful*